537
J159
JACOBWITZ, HENRY
Electricity made simple

APR 8 6

4.95

DATE DUE

MAR 1 3 1988	JAN 1 9 '94	SEP 3 0 199
FEB 2 0 1989	AUG 08 '94	
AUG 1 5 1989	AUG 2 2 94	
AUG 2 8 1989		NOV 0 2 1999
DEC 1 4 1989	FEB 0 9 9	
FEB 1 2 1990	MAR 0 1 199	
MAR 1 0 1990	MAR 1 5 1996	
MAR 05 '9	MAR 2 8 1996	
MAR 1 2 1991	JUN 08 1996	
JUL 1 5 1991	JUN 1 8 1997	
JAN 1 4 1992	JUN 0 2 198	
JAN 0 6 '94	NOV 1 2 99	JAN 2 2 2002

ELECTRICITY MADE SIMPLE

BY

HENRY JACOBOWITZ, B.S.

852979

MADE SIMPLE BOOKS
DOUBLEDAY & COMPANY, INC.
GARDEN CITY, NEW YORK

ABOUT THIS BOOK

Everybody has daily contacts with electricity in various forms and almost everybody has had some more or less formal instruction in the subject. However, most elementary presentations concentrate on the multiplicity of applications of electricity, while brushing lightly over the essential, underlying principles.

The present book does not slight the importance of electrical applications, encountered in all walks of life, but it concentrates its main discussion on the relatively few important principles that make possible the entire electrical industry. The text would be amiss if it did not explain the operation of buzzers and bells, relays and telegraphs, motors and generators, a variety of lamps, and what have you, but the primary emphasis is always on the principles behind the gadgets. Thus, the presentation moves on in a systematic manner from the basic electron theory of matter, through electrostatics, magnetism, electrical current sources, Ohm's Law and direct-current circuits to some basic ideas about electric power and heat. The second half of the book is devoted to the somewhat more difficult concepts of alternating currents, starting with the phenomena of electromagnetism and induction and carrying the presentation through the variety of definitions and practical calculations in alternating-current circuits, impedance and reactance concepts, to a thorough exploration of the meaning of resonance.

By stressing the fundamental laws of electricity the reader not only gains an understanding of electricity in its variety of forms and transformations (in what is hoped to be plain language), but—more importantly—he lays the groundwork for all further studies in advanced electricity, electrical engineering and electronics. (See ELECTRONICS MADE SIMPLE.) Such studies cannot be meaningful without the fundamentals presented in this book.

A word about the experiments. These are meant to illustrate basic principles with a minimum outlay for materials and equipment. Since most experiments use batteries as current source, they are safe, even for youngsters. In the few cases, where the a-c line is used, it would be advisable to place fuses in both sides of the line, and even better, insert a small isolation transformer between the line and the experimental circuit.

Henry Jacobowitz

TABLE OF CONTENTS

CHAPTER ONE

CHAPTER TWO

CHAPTER THREE

CHAPTER FOUR

CHAPTER FIVE

CHAPTER SIX

CHAPTER SEVEN

CHAPTER EIGHT

CHAPTER NINE

CHAPTER TEN

CHAPTER ELEVEN

CHAPTER TWELVE

CHAPTER THIRTEEN

CHAPTER FOURTEEN

THE ELECTRON THEORY

The story of electricity started over 2500 years ago with amusing parlor games, and nothing startling was added to it until the modern era. None of the Greeks who observed the philosopher THALES OF MILETUS (about 600 B.C.) pick up straws and paper with an amber rod that had been rubbed with a cloth, could have suspected that the force behind the toy would become the major means of making man the master of the earth. The Greek word for amber is *elektron* and it was not unnatural, therefore, that the English physicist WILLIAM GILBERT (1540-1603) applied the term "electrics" to materials he found behaving similar to amber. His great treatise *De Magnete,* published in 1600, in which he used such modern terms as *electric force* and *electric attraction* has earned him the title of "father of electricity."

Progress in the next hundred years or so consisted of little more but the observation of isolated electrical and magnetic phenomena. OTTO VON GUERICKE observed in 1660 the sound and light of electrical sparks, which he generated with a crude electric friction machine. The Italian scientist LUIGI GALVANI (1737-1798) noted the twitchings of frogs' legs that were in contact with two dissimilar metals and ascribed them (wrongly) to animal electricity. SIR WILLIAM WATSON (1715-1787) improved the "Leyden-jar" for storing electricity and worked out an early theory of electricity. His experiments and theories were similar to those of BENJAMIN FRANKLIN (1706-1790), the American statesman and scientist, who began his experiments with electricity in about 1746. Franklin developed a practical "condenser" for storing static electricity and first identified lightning with electricity in his famous kite experiment, in 1752. Franklin also developed a coherent **fluid theory** of electricity, but unfortunately guessed wrong about the direction of current flow, which he thought took place from the positive to the negative terminal of a current source. The Franklinian error was not discovered until the present-day **electron theory** had been developed—and by then it had become conventional to describe current as flowing from plus to minus. However, we need not adopt this "conventional" direction of current flow, based on the outmoded Franklinian

theory, but rather shall go straight to the modern electron theory, which has proven highly successful in explaining electrical and magnetic behavior. To appreciate the electron theory, we must first know something about the **atomic structure of matter.**

ATOMIC STRUCTURE

Since the time of the Greeks all matter was thought to be made up of atoms ("atom" is the Greek word for "indivisible"), though the Greek ideas about the nature of these "indivisible" particles were rather vague. It was not till 1802 that the English chemist JOHN DALTON suggested that all matter could be broken down into fundamental constituents, or **elements,** the tiniest particles of which he called **atoms.** There are 98 known elements, of which 92 occur in nature and six are artificially produced in atom smashers and nuclear reactors. Since there are 98 elements, there must be 98 different types of atoms. Through the work of the scientists NIELS BOHR, LORD RUTHERFORD, and others it was revealed that atoms actually have a complex structure, resembling somewhat a miniature solar system. According to Bohr's theory, an atom consists of a central **nucleus** of positive charge around which tiny, negatively charged particles, called **electrons,** revolve in fixed orbits, just as the planets revolve around the sun. In each type of atom, the negative charge of all the orbital electrons just balances the positive charge of the nucleus, thus making the combination electrically **neutral.**

The positively charged nucleus, in turn, reveals a complex structure, but for the purpose of understanding electricity a vastly simplified picture is adequate. According to this simplified picture, the nucleus of the atom is made up of two fundamental particles, known as the **proton** and the **neutron.** The proton is a relatively heavy particle (1840 times heavier than the electron) with a positive (+) charge, while the neutron has about the same mass as the proton, but has no charge at all.

The positive charge on each proton is equal to the negative charge residing on each electron. Since atoms are ordinarily electrically neutral, the num-

ber of positive charges equals the number of negative charges—that is, the number of protons in the
nucleus is equal to the number of electrons revolving around the nucleus. Practically the entire weight
of the atom is made up of the protons and neutrons
in the nucleus, the weight of the orbital electrons
around the nucleus being negligible in comparison.
Lest you should think, however, that substantial
weights are involved, let us at once point out that
the mass of an electron is only about 9.11×10^{-28}
grams (a number with 27 zeros *after* the decimal
point), while that of the proton is only about 1840
times as much, which is still fantastically little. The
proton, on the other hand, is a little smaller than
the electron, having a radius of about 10^{-13} cm. To
give you an idea how small this really is, you might
consider that an electron is about as small compared to a standard ping-pong ball, as a ping-pong
ball is compared to the orbit of the earth, which is
186,000,000 miles in diameter.

TABLE I

Particle	Charge	Weight
Electron	−1	1/1840
Proton	+1	1
Neutron	0	1

Atomic Number. We have seen that all atoms
are made up of protons, neutrons, and electrons.
The difference between various types of elements
results from the number and arrangement of the
protons, neutrons, and electrons within their atoms.
The elements are arranged according to their
atomic number, which is *equal* to the number of
electrons revolving around the nucleus (or to the
number of protons within the nucleus). Thus, an
atom of hydrogen (atomic number 1) has a single
electron spinning around its nucleus, while an atom
of uranium (atomic number 92) has 92 electrons
spinning about the nucleus. The orbits of these
electrons are arranged in "shells" about the nucleus,
each shell having a definite maximum capacity of
electrons. The capacity of successive shells from
the nucleus out is 2, 8, 18, and 32 electrons; however, the *outermost shell contains never more than
eight electrons*. It is this outermost shell which determines the chemical **valence** of an atom and its
principal physical characteristics. The outermost
shell is also the most important for electricity, since
it is the only one from which electrons are relatively easily dislodged to become "free" electrons
capable of carrying a current in a conductor. The
electrons in the inner shells cannot be forced out

easily from their orbits and, hence, are said to be
"bound" to the atom.

Atomic Weight. The weight of an atom, called
atomic weight, is determined almost entirely by the
sum of the number of protons and neutrons within
its nucleus. Atomic weights are *relative*, since they
do not state the number of ounces each atom
weighs, but compare the weight of one atom with
that of another. Thus, the atomic weight of oxygen
is 16, that of helium is 4, while that of hydrogen is
1. Hence, an oxygen atom is 16 times as heavy as a
hydrogen atom and four times as heavy as a helium
atom. You can determine the number of neutrons
in an atom by *subtracting* the atomic number
(equals the number of protons) from the atomic
weight (equals the number of protons + neutrons).

TABLE II

Atomic Property	Explanation
Neutral Atom	Number of electrons = number of protons
Atomic Number	Number of electrons = number of protons = atomic number
Atomic Weight	Number of protons + number of neutrons = atomic weight
No. of Neutrons	Atomic weight − atomic number = number of neutrons

Let us look at some examples of our atomic
model. (See Fig. 1.) The top illustration shows an
atom of hydrogen, while the bottom illustration
shows the more complex structure of an atom of
carbon. The hydrogen atom has only one proton in
its nucleus (charge +1), surrounded by a solitary
orbital electron (symbolized e−). It has no neutrons
at all. Consequently, its atomic number is 1 and its
atomic weight is also 1. It is the simplest atom of
all the elements. The carbon atom, which looks a
little more like a miniature solar system, has a nucleus containing six neutrons and six protons (total
charge +6). Its atomic weight, therefore, is 12
(number of protons + number of neutrons). To balance the positive charge of the nucleus, the carbon
atom is surrounded by six orbital electrons (6 e−),
two of which are in the innermost shell, while the
other four are in the outer shell.

The most complex atom occurring in nature is
uranium, which has an atomic weight of 238 and
atomic number 92. Thus, the uranium atom must
have 92 orbital electrons and a nucleus containing
92 protons and 146 neutrons. It is a little difficult
to draw a picture of it. Still more complex atoms
have been created artificially in the cyclotron and

HYDROGEN

At. No. 1

At. Wt. 1

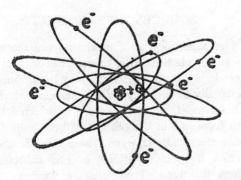

CARBON

At. No. 6

At. Wt. 12

Fig. 1. Structure of Hydrogen and Carbon Atoms

nuclear reactors, but these are generally unstable and break down into lighter elements.

Molecules. While atoms are the smallest bits of matter in each element, it may be well to keep in mind that most materials in the world are compounds of various elements, formed by combinations of different atoms. These smallest combinations of atoms are called molecules.

IONS AND IONIZATION

An ion is an atom (or molecule) that has become electrically unbalanced by the loss or gain of one or more electrons. An atom that has lost an electron is called a positive ion, while an atom that has gained an electron is known as a negative ion. The reason is clear. When an atom loses an electron, its remaining orbital electrons no longer balance the positive charge of the nucleus, and the atom acquires a charge of +1. Similarly, when an atom gains an electron in some way, it acquires an excess negative

charge of −1. The process of producing ions is called ionization.

Ionization does not change the chemical properties of an atom, but it does produce an electrical change. It can be brought about in a number of ways. As we have seen, the electrons in the outermost shell of an atom are held rather loosely and, hence, can be dislodged entirely by collision with another electron or atom, or by exposure to X rays. Ionization is important in electron tubes.

FREE ELECTRONS

Electrons that have become dislodged from the outer shell of an atom are known as free electrons. These electrons can exist by themselves outside of the atom, and it is these free electrons which are responsible for most electrical and electronic phenomena. Free electrons carry the current in ordinary conductors (wires), as well as in all types of electron tubes. The motion of free electrons in antennas gives rise to electromagnetic radiations (radio waves).

Conductors and Insulators. Most substances normally contain a number of these free electrons that are capable of moving freely from atom to atom. Metallic materials, such as silver, copper, or aluminum, which contain relatively many free electrons capable of carrying an electric current, are called conductors; non-metallic materials, which contain relatively few free electrons, are called insulators. Materials that have an intermediate number of free electrons available are classed as semiconductors. Actually, there are no perfect conductors and no perfect insulators. The more free electrons a material contains, the better it will conduct. All substances can be arranged in a conductivity series, in accordance with their relative number of free electrons available.

Electric Current. The free electrons in a conductor are ordinarily in a state of chaotic motion in all possible directions. But when an electromotive force (emf), such as that provided by a battery, is connected across a conductor, the free electrons are guided in an orderly fashion, atom to atom, from the negative terminal of the battery, through the wire, to the positive terminal of the battery. (See Fig. 2.) This orderly drifting motion of free electrons under the application of an electromotive force (or voltage) constitutes an electric current. Although the electrons drift through the wire at a relatively low speed, the disturbance or impulse is transmitted almost at the speed of light. Note that the electron current continues to flow only as long as the wire remains connected to the battery. The wire conductor itself remains electrically neutral, since electrons are neither gained nor lost by the atoms within the wire. What happens is this: Electrons enter the wire from the negative terminal of the battery and an *equal* number of electrons is given up by the other end of the wire to the positive battery terminal. Thus, the free electrons present within the wire act simply as **current carriers**, which are continually being replaced, but none are lost.

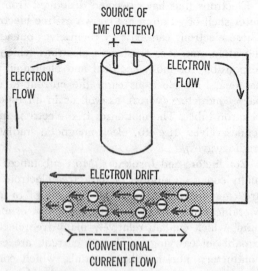

Fig. 2. Conduction of Electricity Through a Metallic Conductor

Note also in Fig. 2 that the "conventional" or "Franklinian" current is from the **positive to the negative terminal** of the battery, in a direction *opposite* to the electron flow. Such a current would involve the transfer of **positive** charges from the + to the − terminal of the battery, which actually does *not* take place. Conventional current is still widely used in the markings of meters, formulation of electrical rules, and in many text books. We shall not use it, however, and the term "current" from now on shall always designate electron flow from minus (−) to plus (+).

Resistance. Since an electric current is the flow of free electrons, materials that have a large number of available free electrons permit a greater current flow for a given applied electromotive force (voltage) than do materials with relatively few free electrons. The measure of the opposition to the flow of free electrons in a material is a quantity called

resistivity. As we shall see later in more detail, the **resistance** to the flow of electric current of a certain material of given cross section and length can be calculated from its resistivity. As mechanical friction, the resistance of a material **dissipates energy** in the form of **heat** because of collisions occurring between free electrons and atoms. Conversely, if a material of a given resistance is **heated**, more collisions take place and the resistance to the flow of electric current **increases**.

SOURCES OF ELECTRICITY

The chief sources of electricity are **mechanical, chemical, photoelectric, thermoelectric,** and **piezoelectric** in nature. Electricity may be produced **mechanically** in two ways. When certain materials are rubbed together, electrons are transferred by friction from one to the other, and both materials become **electrically charged.** These charges are not in motion, but reside **statically** on each substance and hence this type of electricity is known as **static electricity** or **electrostatics.** Electricity may also be generated mechanically by the relative **motion** of a conductor with respect to a **magnetic field,** a process known as **induction.** The interaction of electric and magnetic fields is studied in a branch of electricity called **electromagnetism.** Practically all commercial electricity is produced by electromagnetic generators.

Electricity can be generated **chemically** by inserting two dissimilar metals, such as zinc and copper, into a conducting solution called **electrolyte.** An electromotive force (emf), or **voltage,** is then found between the metals, which can cause current to flow through an externally connected conducting circuit. By connecting a number of such chemical **cells** together as a **battery** any desired voltage and quantity of electricity can be supplied. Electricity produced by chemical action is studied in **electrochemistry.**

Sunlight or artificial illumination falling upon certain **photosensitive** materials, such as cesium or selenium, produces electricity by knocking out free electrons from the surface of the material. This process is known as **photoelectric emission,** or simply **photoelectricity.**

When the junction of two dissimilar metals, such as an iron wire welded to a copper wire, is heated, an electromotive force (emf) appears between the free ends of the metals. Such a junction is called a thermocouple and the process is termed thermo-electricity.

Electricity, finally, may be generated by the mechanical compression, stretching and twisting of certain crystals, such as quartz and Rochelle salts. Materials that permit generating an emf by mechanical pressure are called **piezoelectric** and the process is known as **piezoelectricity**.

Historically, electricity was first observed in static form, as in the electrification of amber by rubbing and the lightning discharges of electricity by charged clouds. In the chapters that follow we shall study the action of electricity in all these currently known forms.

Practice Exercise No. 1

1. Explain the origin of electricity and some high points in its early development.

2. How many elements exist in Nature? How many types of atoms? What is the *total* number of elements and atoms presently known?

3. Explain Bohr's concept of atomic structure.

4. Make a brief table listing the charge and relative weight of the three main types of atomic particles.

5. Draw the atomic structure of hydrogen and carbon, and assign the proper atomic number and atomic weight to each.

6. The element *neon* (an inert gas) has atomic number 10 and an approximate atomic weight of 20. How many neutrons are contained in the neon nucleus? Can you draw a sketch of the atomic structure of neon, showing the number of electrons in each shell? Can you guess why the element is *inert* (i.e., forms no *compounds* with other elements)?

7. Explain the process of *ionization* and how positive and negative *ions* may be produced. What might *double* ionization mean?

8. What are *free electrons* and how are they produced?

9. Distinguish between conductors, semiconductors, and insulators on the basis of the electron theory.

10. What constitutes an electric current and what is its direction?

11. Explain why a wire through which an electric current is flowing does not become electrically charged or remain charged after the current flow stops.

12. Explain qualitatively the action of *resistivity* and *resistance* and the factors which affect them. Why is heat produced and what is the effect of heating a conductor?

13. List the five main sources of electricity and explain the basic action involved in each. State the proper term for each process and the branch of electricity concerned with it.

SUMMARY

The Greek philosopher Thales observed **static electricity** (amber charged by rubbing) in about 600 B.C.

Benjamin Franklin developed an early **fluid theory** of electricity, in which he mistakenly assumed the flow of (positive) electric current from plus (+) to minus (−).

The modern explanation of electricity is by means of the **electron theory**, which is based upon the **atomic structure** of matter.

There are 98 known **elements**, corresponding to 98 different types of atoms. Of these 92 occur in Nature, while the remaining six are artificially created in atom smashers and nuclear reactors.

An **atom** is the smallest particle of an element that shows its chemical and physical properties.

Atoms resemble **miniature solar systems**, consisting of a central **nucleus** of **positive** charge, around which tiny, **negatively** charged particles, called **electrons**, revolve in fixed orbits. The negative charge of all orbital electrons just *balances* the positive charge of the nucleus and, hence, the atom is electrically **neutral**.

The nucleus of the atom is made up of **protons** and **neutrons**. The proton is 1840 times heavier than the electron and has a **positive charge**. The neutron has the same mass as the proton, but has **no charge** at all.

Electron orbits are arranged in shells about the nucleus, with capacities of 2, 8, 18, and 32 electrons (from the nucleus out). The outermost shell of an atom cannot contain more than eight electrons.

Atomic number refers to the total number of electrons in the shells or to the total number of protons in the nucleus.

The **atomic weight** of an atom is the sum of the number of protons and neutrons in the nucleus. The number of neutrons equals the **difference** between the atomic weight and atomic number.

An atom that has lost an electron is called a **positive ion**; one that has gained an electron is called a **negative ion**. **Ionization** is usually produced by collisions between atoms and electrons.

Free electrons are electrons dislodged from the outer shell of an atom. They may exist by themselves and can act as carriers of electricity in conductors or vacuum tubes.

Conductors contain relatively many free electrons, **insulators** relatively few; **semiconductors** have an intermediate number of free electrons.

An **electric current** is an orderly drifting motion of free electrons in a conductor under the influence of an applied **electromotive force** (emf), or **voltage**. The **direction** of electron motion is from the **negative** terminal of the current source to the **positive** terminal. **Conventional** or **Franklinian** current flows in the opposite direction.

The opposition to the flow of electric current or **resistivity** of a material depends on the relative number of available free electrons.

The **resistance** of a conductor depends on its resistivity, its cross section and its length. Resistance produces heat because of collisions between free electrons and atoms. When a conductor is heated, its resistance **increases**.

The chief sources of electricity are: 1. **mechanical**, by **friction** (electrostatics) or by motion of a conductor with respect to a **magnetic field** (electromagnetism); 2. **chemical**, by the insertion of two dissimilar metals in a conducting solution or **electrolyte**; 3. **photoelectric**, by light falling upon a **photosensitive** surface; 4. **thermoelectric**, by the heating of a junction of two dissimilar metals (**thermocouple**); and 5. **piezoelectric**, by mechanical pressure applied to certain **crystals** (quartz, Rochelle salts).

CHAPTER TWO

ELECTROSTATICS—CHARGES AT REST

Let us now look at the kind of electricity—electrostatics—with which the ancients used to amuse themselves by **electrifying** substances through rubbing. We've all had experience with **static** electricity: lightning during a thunderstorm; sparks flying after we shuffle over a deep-pile rug; hair standing up on end after vigorous combing or brushing—all these are typical examples of the effects of static electricity. The term **electrostatics**, which refers to electricity at rest, is something of a misnomer, since we now know that the carriers of electricity—the electrons—are in continual motion. The term is still useful, however, to distinguish between the **random** motion of electrons residing on the surface of a charged (electrified) body and the **orderly** drifting motion of electrons taking place when an electric current flows through a conductor.

Charging by Contact. Any substance, under suitable conditions, can become **electrified** or **charged** to some degree. A glass rod, rubbed with silk or a hard-rubber rod rubbed with fur, becomes charged and attracts little pieces of paper. A sheet of paper, when rubbed vigorously becomes charged and clings to a wall. A simple experiment shows that there are two different kinds of electrification or charge.

EXPERIMENT 1: Suspend two *pith balls* (light, soft balls covered by conductive aluminum paint) from dry silk or nylon threads a couple of inches apart from each other, as shown in Fig. 3. Touch each of the balls with a glass rod that has been charged by rubbing it with silk. Note that each of the balls is initially attracted by the charged rod, but is repelled by it after making contact and acquiring some of its charge. Moreover, the two charged pith balls repel each other and remain separated as long as the charge remains on them. (Note: the experiment generally succeeds only during clear, dry weather. On a moist day it may be necessary to keep all materials in a hot, dry place, such as a drying oven.)

Discharge the balls by touching them with your finger or wait till the original charge has leaked off. Now charge the balls again by touching each with a hard-rubber (vulcanite) rod charged by rubbing with fur or catskin. (The same result can be obtained by using a rod of ebonite or sealing wax rubbed with catskin or flannel; plastics, such as polystyrene or lucite, work well even on humid days.) Note that the same thing happens as before: the balls are initially attracted to the rod, but after contact they are repelled by the rod, as well as by each other. Again, discharge the balls by touching them.

Now charge one of the balls by touching it with the charged glass rod and charge the other by bringing it in contact with the charged rod of rubber or sealing wax. Note the radically changed behavior in this case. While each of the balls is initially attracted to its respective rod and repelled from it after contact, as before, the *two balls now attract each other and cling together after having*

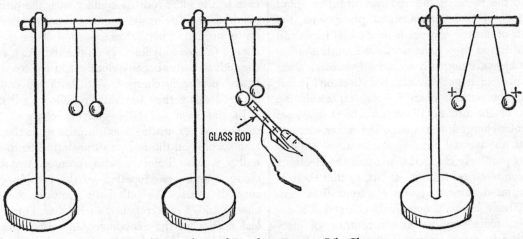

GLASS ROD

Fig. 3. Electrical Repulsion Between Like Charges

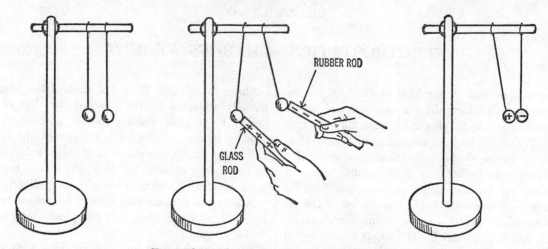

RUBBER ROD

GLASS ROD

Fig. 4. Electrical Attraction Between Unlike Charges

been charged. The balls will continue to attract each other until their charge has leaked off.

The experiment demonstrates clearly that the electricity on the glass rod differs from that on the hard-rubber rod or sealing wax. Benjamin Franklin, quite arbitrarily, called the charge acquired by the glass rod when rubbing it with silk *positive* electricity, while he assigned a *negative* charge to the hard-rubber rod or sealing wax, when rubbed with fur or flannel. It is further evident from the experiment that the pith balls *repel* each other when they are charged alike either (+ or −), and *attract* each other when they are charged *oppositely* (+ and −). The experiment thus confirms a fundamental fact of electricity: **like charges of electricity repel each other, and unlike charges attract each other.** This important fact was discovered by the French chemist CHARLES DU FAY in 1733.

While the theorists differed for centuries about the nature of these fundamental phenomena, the behavior of electric charges is now easily explained by the electron theory. As we have seen the atoms of any object are normally electrically **neutral**, since the number of negatively charged electrons is just equal to the positive charges (protons) within the nucleus of the atoms. When an object becomes electrically **charged**, it has acquired either more or less than the normal number of electrons. A body becomes **positively** charged if some of the electrons have been removed from its atoms, so that there is an **electron deficiency** (fewer electrons than protons). A body becomes **negatively** charged if it acquires—in some way—an **excess** number of electrons; that is, more electrons than protons. Thus, when a glass rod is rubbed with silk, some of the

loosely held electrons on the surface of the rod are detached and transferred to the silk. The glass rod, consequently, loses electrons and becomes **positively** charged, while the silk cloth gains a surplus of electrons and becomes **negatively** charged. Similarly, when the hard-rubber rod (or similar substance) is rubbed with fur or flannel, the friction "strips" some of the electrons near the surface of the fur or flannel cloth and transfers them to the hard-rubber rod. As a result, the hard-rubber rod acquires a surplus of electrons and becomes negatively charged, while the fur or flannel cloth is left with a deficiency of electrons and becomes positively charged.

In the experiment, the conducting pith balls were *repelled* from either the positive glass rod or the negative hard-rubber rod, after being brought *in contact* with it. (We shall presently explain why they were *initially* attracted before contact.) In the case of the glass rod the contact with the pith ball resulted in drawing off some of the free electrons on its surface to neutralize a *portion* of the positive charge (electron deficiency) on the rod. As a result, the pith ball also became deficient in electrons and, hence, *positively* charged. With both rod and ball charged alike, they naturally repelled each other.

In the case of the negatively charged hard-rubber rod (or sealing wax), a portion of the electron surplus on the rod was transferred to the pith balls, which therefore also became negatively charged and were repelled by the rod. In either case, of course, the pith balls acquired the same charge and, hence, repelled each other. During the last portion of the experiment, in contrast, one of the balls was charged positively by contact with the glass rod, while the other was charged negatively by contact with the hard-rubber rod or sealing wax. Having acquired **opposite** charges, the two balls **attracted** each other.

Charging by Induction. Let us now return to the question why the neutral pith balls were initially attracted to the charged rods before making contact with them. To explore this question experimentally, let us construct a more sensitive electrical charge detector than the pith balls, known as the **leaf electroscope**. The best type contains two thin gold leaves at the bottom of a metal rod, but to save expense we shall make one using aluminum leaves.

EXPERIMENT 2: Construct the aluminum leaf electroscope, shown in Fig. 5, from a large jar and lid, a cork, copper (bell) wire, a copper washer or disk (new penny) and some aluminum foil. Drill a hole in the lid of the jar to receive the cork. Pierce cork to make a hole and work the copper wire through. Bend the lower end of the wire into T-shape, as shown in (a) of Fig. 5. Fasten the upper end of the wire to the copper disk, either by soldering it or by winding it through two holes drilled into the disk. Use nail polish or plastic cement to seal the juncture of the cork and lid. Now remove a piece of aluminum foil from a cigarette pack or gum wrapper, using alcohol to separate the thin foil from its paper base. Cut two strips of foil, each about an inch long and one-half inch wide; mount them through the T-shaped end of the copper wire and glue with nail polish or plastic cement. (See Fig. 5b.) For best results, the jar and lid assembly should be heated in an oven for a few minutes to drive out all moisture. Be sure to screw the lid on tightly before it cools off. This completes construction of the electroscope.

We are now ready to try some experiments with our home-made electroscope. First rub a glass rod

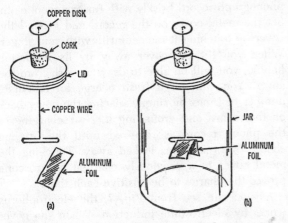

COPPER DISK

CORK

LID

COPPER WIRE

JAR

ALUMINUM FOIL

ALUMINUM FOIL

(a)

(b)

Fig. 5. Constructing an Aluminum Foil Electroscope

with silk and bring the charged rod in contact with the copper disk on top of the electroscope. Note that the aluminum leaves will immediately fly apart, indicating that both have been charged by the glass rod with the same polarity, in this case positive. If you now touch the copper disk with a rod of hard-rubber or sealing wax that has been rubbed with flannel, the leaves of the electroscope will partially collapse, indicating that the charge on the rubber rod is of opposite sign—that is, negative. You can cause the leaves to collapse completely by touching the copper disk with your finger. Your finger will act as conductor, discharging the electroscope by leading the charges to the floor (ground).

Next try the following experiment. Charge a rod of hard rubber, sealing wax or ebonite and approach the disk of the neutral electroscope closely, but do not touch it. The leaves of the electroscope will fly apart (Fig. 6b), though there is no contact between the rod and copper disk. What happens is this: When the negatively charged rod approaches the copper disk, a **redistribution** of free electrons and positive copper nuclei takes place within the disk. The positive nuclei of the copper atoms are attracted toward the rubber rod and shift toward that end of the disk, while the (negative) free electrons are repelled along the copper wire into the aluminum leaves. The leaves, consequently, diverge, having been negatively charged. Since the positive charges (copper nuclei) are held fixed in position on the copper disk by the rubber rod, this charge is referred to as a **bound induced** charge. The negative charge on the aluminum leaves, in contrast, is called the **free induced** charge. These induced charges are only temporary, however. If you move the rubber rod away from the electroscope, the leaves will collapse, indicating that all charges have redistributed themselves again into their normal, neutral configuration. Note that this temporary induced charge also explains the momentary attraction of the pith ball pendulum to a charged rod, before making contact with it (Experiment 1).

You can make the induced charge permanent by simply leading off the free charges to "ground." To do this, approach the electroscope again with the charged hard-rubber (or ebonite) rod until the leaves are seen to diverge. Now "ground" the disk of the electroscope by touching it with your finger, while *still keeping the rod at the same distance* from the disk. Note that the leaves collapse (Fig. 6c). This is caused by the "leaking off" of the free

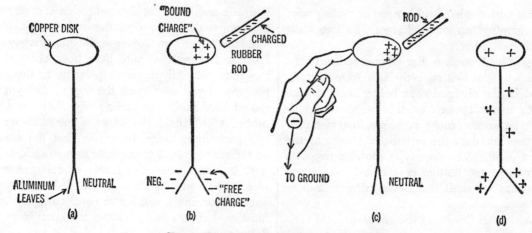

Fig. 6. Charging an Electroscope by Induction

electrons to ground through your body, while the "bound" charge is still held by the attraction of the rod. If you now remove the rubber rod, the bound, positive charge will be liberated and will distribute itself throughout the conducting parts of the electroscope, causing the leaves to diverge again. The leaves are now permanently charged **positive** by induction (Fig. 6d). You can check the *sign* of the charge by touching the electroscope first with a charged glass rod and then with a charged hard-rubber or ebonite rod. When touched with the glass rod, the leaves will be seen to diverge even more strongly, indicating that the polarity of their charge is the same as that of the glass rod, or **positive.** Touching the electroscope with the rubber rod, however, will cause the leaves to *collapse*, indicating that the charge on the leaves is of **opposite** sign than the negative charge on the rubber rod. Again, this proves conclusively that the electroscope has been charged **positively** by induction from the **negative** rubber rod.

If you repeated the entire experiment by using a (positively charged) glass rod to charge the electroscope by induction, in place of the hard-rubber rod, you would make exactly the same observations, *except* that the final charge residing on the leaves would turn out **negative.** We conclude, therefore, that **the charge induced is always of a polarity opposite to that of the inducing body.** In contrast, when the electroscope is charged *by contact* with another charged body, the charge on the leaves is of the *same sign* as that of the charging body, as we have seen.

The Electrophorus. An interesting application of charging by induction is the **electrophorus**, one of the oldest electrostatic generators (see Fig. 7). It consists of a shallow metal pan into which a cake of

sealing wax or resinous material has been melted; and a metal disk of smaller diameter than the pan, provided with an insulating handle fits on top of the cake. By rubbing the sealing wax with fur or wool, placing the disk on top of the cake and grounding it, a **positive** charge is induced in the disk, as can be verified with an electroscope. Each time the disk is placed on top of the sealing wax, grounded, and then lifted away, an additional positive charge appears on the disk, without the need for recharging the wax. Thus, an **unlimited** number of electrostatic charges can be obtained from a single charging of the cake of sealing wax. Let us try this out.

EXPERIMENT 3: Construct an electrophorus by placing a phonograph record (shellac or vinylite LP) snugly in a metal pie plate. Place a flat metal cover, somewhat smaller than the pie plate and provided with an insulated handle, on top of the record and you have an elementary electrophorus (Fig. 7).

Now charge the electrophorus by rubbing the phonograph record briskly with fur or a wool cloth. Set the metal cover on the record and ground the cover by touching it momentarily with your finger. When you lift the cover away by its insulating handle, you will be able to draw a spark from the cover. You can get a fresh charge an indefinite number of times by simply placing the cover back on the record and grounding it. No recharging of the phono record is required until the original charge eventually has leaked away. Touching the metal cover to a previously charged electroscope proves the charge to be **positive** each time.

As you can see from Fig. 7, the electrophorus works by electrostatic induction. When the cover is placed on the negatively charged phonograph

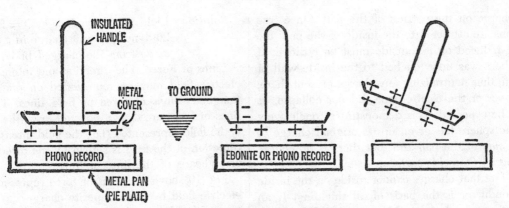

Fig. 7. Action of the Electrophorus

disk (or sealing wax), a bound, positive charge is induced at the bottom of the metal cover and a free, negative charge at its top. Grounding the cover by touching it leads the negative charge away, so that only the positive charge remains. Lifting the cover away from the disk, against the electrostatic attraction, "frees" the positive charge and makes it available. The energy in the unlimited number of charges is accounted for by the work you do each time, when you lift the cover from the disk against the force of electrostatic attraction. You will be surprised to observe the considerable strength of the force between the charges.

Van de Graaff Electrostatic Generator. There are a number of continuously operating **electrostatic generators**, which produce large amounts of electric charge at very high voltages, for lightning studies and atom smashing. One of these is the Van de Graaff generator, which produces potentials of several *million* volts by transporting electrostatic charges from a continuously moving belt to a large hollow sphere (Fig. 8). As shown in the diagram,

negative electric charge is sprayed onto an endless fabric belt (silk or linen) between a comb of needle points (*A*) and a rounded surface (*B*). The negative charge is mechanically transported by the pulley-driven belt to another comb-shaped collector (*D*), which transfers it to the outside of the metal dome. Inasmuch as the process is continuous and the belt can be run at high speed, enormous charges can be built up on the dome, amounting to millions of volts. The dome is supported by an insulating column and the maximum voltage is limited only by direct electrical discharge from the metal shell and the quality of the insulation.

Location of Charge. The English scientist MICHAEL FARADAY was the first to prove that no electric charge can be stored *within* a hollow conductor. All charges reside on the *outside* of the conducting surface. In his famous ice-pail experiment, he lowered a small, positively charged metal sphere into a metal ice pail by means of an insulating thread (Fig. 9). The outside of the pail was connected with a wire to a leaf electroscope. As soon as the sphere was inserted into the pail, the leaves of the electroscope diverged, proving the presence

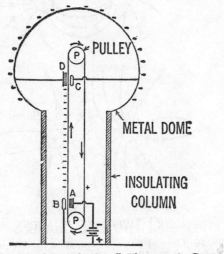

Fig. 8. Diagram of Van de Graaff Electrostatic Generator

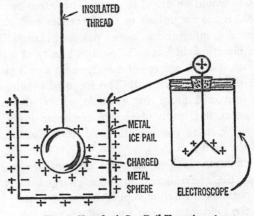

Fig. 9. Faraday's Ice-Pail Experiment

of a charge on the outside of the pail. Since free electrons are attracted to the inside of the pail, the charge induced on its outside must be **positive**. If the sphere was now touched to the inside wall of the pail, thus neutralizing any charge present there, the leaves of the electroscope did *not* collapse, as would be expected. This demonstrates conclusively that the sphere has given up its original charge to the *outside* of the pail and that there is *no charge residing on its inside surface.*

The fact that charges cannot reside on the inside of a conductor is the basis of all **shielding**. If an electroscope is placed inside a shield of fine wire mesh, its leaves will not diverge, even when large electrostatic charges are placed on it. A radio placed inside such a wire shield will not play. This also accounts for the comparative safety from lightning of passengers inside an automobile or metal airplane. Tenants inside a steel-frame building are protected against severe thunderstorms, even without a lightning arrester. Lightning itself is caused, of course, by electric discharges between *charged* clouds and the earth's surface.

ELECTRIC FIELD OF FORCE

As we have seen, an electrically charged body exerts a physical force on other bodies around it and can make them do work by moving or lifting them. (See Fig. 3.) The area of influence in the vicinity of such a charged body is known as an **electric field of force**, or simply an **electric field**. Since this field is capable of performing work, it is endowed with energy. If an electrically charged test body is inserted into this field, it will be either attracted or repelled, depending on the sign of its charge. The *direction* of an electric field at any point is the direction in which a **positively** charged body would be urged to move. The **intensity** of the field is the **magnitude of the force per unit charge**, and it is measured in dynes per unit charge. Thus, an electric field has an intensity of unity at a point, when it exerts a force of **one dyne on a unit charge**. (A dyne is a small unit. The force of gravity on a weight of 1 gram, for example, is 980 dynes.) In general, if Q units of charge are acted upon by a force F in an electrostatic field, then the intensity of the field E is F/Q.

EXAMPLE 1: Ten unit test charges are inserted into the electric field about a charged body and experience a force of repulsion of 200 dynes. What is the field intensity at that point?

Solution: Field intensity $E = F/Q = 200/10 =$ 20 dynes per unit charge in a direction *away* from the charged body.

Lines of Force. The direction and intensity of an electric field may be represented on diagrams by imaginary **lines of force** or **field lines**. The more lines of force are drawn *per unit area*, the stronger is the field represented (i.e., the field intensity). The direction of the field is shown by the direction and arrowheads of the lines of force.

Fig. 10 shows the lines of force representing the electric field between opposite charges (a) and between like charges (b). A small, **positive** test charge would tend to move in the direction of the field lines. If such a test charge were inserted into the field between like charges (Fig. 10*b*), it would be repelled toward the center and probably squeezed out sideways, as shown by the **repelling** lines of force. When inserted into the field between *unlike* charges (Fig. 10*a*), however, the test charge would follow the lines of force between + and −, and would be strongly attracted toward the **negative** charge. Although the illustration (Fig. 10) shows only the lines of force in the plane of the page, you must imagine the field lines surrounding the charged bodies in all three dimensions of space.

You can easily verify that an electric field of force actually exists around charged bodies with the intensity and directions indicated by the lines of force. For instance, if you were to place two highly charged bodies (metal disks or balls) on a

(a) TWO UNLIKE CHARGES

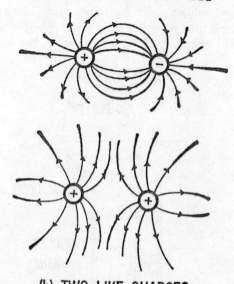

(b) TWO LIKE CHARGES

Fig. 10. Lines of Force Representing the Electric Field Between Unlike Charges (a) and Between Like Charges (b)

glass plate and scattered some cigarette ashes or fine cork filings between them, you would observe the particles arrange themselves in a pattern indicated by the lines of force in Fig. 10a or b, depending on the polarities of the charges.

Coulomb's Law of Force. Our experiments have shown that like charges repel and unlike charges attract each other, but we do not know the **magnitude** of the force of attraction or repulsion. To establish the magnitude of the force, the French physicist CHARLES A. DE COULOMB (1736-1806) made a series of quantitative measurements of the forces between two charges by means of a **torsion balance.** Using this device with varying charges and distances, and in different surroundings, Coulomb was able to show in 1785 that **the force between concentrated (point) charges varies directly with the product of the individual charges, and inversely with the square of the distance between them.** As we shall see in the next chapter, this law of force also holds true for magnetic fields, and as a matter of fact, the **inverse square law** is valid for practically all fields of force, including gravitation.

Coulomb also established that the magnitude of the force is the *same* for attraction or repulsion between the charges. If the charges are alike (both $+$ or both $-$), the force will be one of repulsion, while for unlike charges the force is one of attraction. Moreover, it became evident that the force between the charges was **influenced by the medium** in which the charges were placed. All these relationships are summarized in simple form by **Coulomb's Law of Force.** According to his law, the force (F) between two *point* charges, q_1 and q_2, is

$$F = \frac{q_1\, q_2}{k\, r^2}$$

where r is the distance between the charges and k is called the **dielectric constant** of the medium. This constant is taken as *unity* for a vacuum and it is nearly so for air at normal pressure and temperature ($k = 1.000586$). For glass k varies from 4 to 8, for paper it is 2.5, for quartz 4.5, etc. The relations expressed by Coulomb's law are shown in schematic form in Fig. 11 for two point charges (in vacuum).

Once we have chosen appropriate *units* for the force (F) and the distance (r), Coulomb's Law serves to define the **unit charge** we have previously mentioned. An **electrostatic unit charge** (abbrevi-

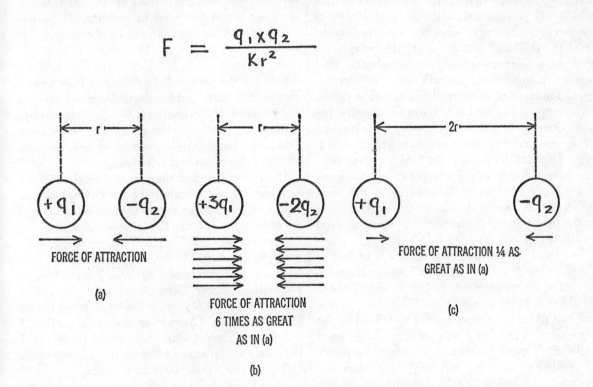

Fig. 11. Coulomb's Law for two Point Charges

ated **esu**) is defined as a charge which will repel another unit charge of the same sign with a force of *one dyne,* when the distance between the charges (in a vacuum) is *one centimeter* (2.54 cm = 1 inch). The electrostatic unit charge (esu), which is sometimes referred to as the **statcoulomb,** is a very small unit. A much larger unit, called the **coulomb,** is the equivalent of three billion (3×10^9) esu. A coulomb of charge will exert a force of 9×10^9 NEWTONS, or 9×10^4 dynes, on another coulomb of charge, when separated from it by a distance of *one meter* (39.37 inches). An example will clarify the use of the units.

EXAMPLE 2: Compute the force of repulsion in air (assume k = 1) between a point charge of +30 esu and a point charge of +20 esu, for a distance of 10 cm between the charges.

Solution: $F = \dfrac{q_1\, q_2}{k\, r^2} = \dfrac{+30 \times +20}{1 \times (10)^2} = \dfrac{600}{100} = 6$ dynes.

Electric Potential. If you life an object from the floor and place it on a shelf, you will have performed a certain amount of work equal to the weight of the object (force of gravity) times the distance (height) you have lifted it. This work is stored in the object in the form of **potential energy,** which will be returned in the form of energy of motion (kinetic energy), when the object is dropped from the shelf and hits the floor. Similarly, when you move a charged body in an electric field against its *opposition,* you will have performed a certain amount of work, which is stored as potential energy. If the field is *uniform* in intensity, the work done (potential energy stored) is the product of the constant force and the distance the charge is moved against the force. If the field is not uniform, the force varies from point to point, and the work is not easily determined. In either case, the work performed in moving the charge to a certain point in the field is **equal to the potential energy stored** by the charge.

In electricity, the potential energy is known as the electric **potential** and its significance is the same as that of the mechanical **level** or height to which an object is lifted. Accordingly, the electric **potential at a point in an electric field is defined as the work done in moving a unit positive charge from an infinitely great distance to the particular point.** (The distance must be "infinitely great" since the influence of an electric field theoretically extends to infinity, though in practice its effects are measurable only over a certain range.)

The potential (symbol V) at a point in an electric field is said to be **one volt** if one **joule** of work (0.737 ft-lbs) must be performed to bring **one coulomb** of charge from infinity to the point in question. The potential (V) thus expresses the work (in joules) per unit charge (coulomb) transferred. A potential of 10 volts, hence, represents an expenditure of energy of 10 joules per coulomb, 50 volts is the equivalent of 50 joules energy per coulomb, and so on.

Potential at a Point. The potential at a point in an electric field is a measure of the work that must be done to bring a unit positive charge from infinity to the point in question. The *greater* the charge that is responsible for the electric field, the more work must be done to bring the unit charge to the point against the repelling effect of the field and, hence, the greater is the potential. The potential in the vicinity of a concentrated (point) charge, thus, is directly *proportional* to the *amount* of the charge (Q). Further, the *closer* the unit charge must be brought to the repelling field of the point charge, the more work must be done, and, consequently, the greater is the potential. The potential at a point near a concentrated charge is therefore **inversely proportional to the distance** (r) between the charge and the point in question. These two relations may be summarized by the simple formula:

$$V = \frac{Q}{k\, r}$$

where V is the potential at a distance r from a point charge of Q units, and k is the dielectric constant of the medium. If the charge is expressed in electrostatic units (esu) and the distance in centimeters (cm), the potential (V) will be in **ergs per unit charge** (also sometimes called **esu**).

If you want to find the potential in the vicinity of a number of concentrated point charges, simply add up the potentials due to each separate charge. An example will clarify the procedure.

EXAMPLE 4 (Fig. 12): Three charges of +7, +49, and +21 esu, respectively, are placed at the north, west and east points of a circle of 7 cm radius, as shown in the figure. Compute the total potential at the center (point D) and at the south point (point E) of the circle. (Assume vacuum as a medium.)

Solution: The potential at the center (point D) of the circle is the sum of the individual potentials Q/kr, where r equals the radius. Hence,

$$V_D = Q_A/kr + Q_B/kr + Q_C/kr =$$

$$\frac{7}{1 \times 7} + \frac{49}{1 \times 7} + \frac{21}{1 \times 7} = 11 \; esu$$

The potential at the south point (point E) may be determined by computing first the distances from the charges to point E. The distance from Q_A to point E is equal to the diameter, or 14 cm. The distance from either Q_B or Q_C to point E is the length of the hypotenuse of a right triangle, formed by the radii of 7 cm each. This distance, thus, is $\sqrt{7^2 + 7^2} = \sqrt{98} = 9.9$ cm. The potential at point E (V_E), hence, is

$$V_E = \frac{7}{1 \times 14} + \frac{49}{1 \times 9.9} + \frac{21}{1 \times 9.9} = 7.57 \; esu.$$

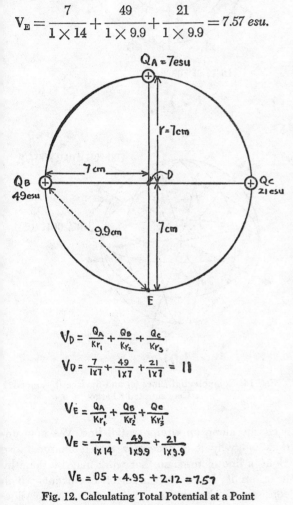

$$V_D = \frac{Q_A}{Kr_1} + \frac{Q_B}{Kr_2} + \frac{Q_C}{Kr_3}$$

$$V_D = \frac{7}{1 \times 7} + \frac{49}{1 \times 7} + \frac{21}{1 \times 7} = 11$$

$$V_E = \frac{Q_A}{Kr_1'} + \frac{Q_B}{Kr_2'} + \frac{Q_C}{Kr_3'}$$

$$V_E = \frac{7}{1 \times 14} + \frac{49}{1 \times 9.9} + \frac{21}{1 \times 9.9}$$

$$V_E = 0.5 + 4.95 + 2.12 = 7.57$$

Fig. 12. Calculating Total Potential at a Point

Potential Difference. The **absolute** potential at a point is rarely important in practice. We are usually interested in the work performed in moving a unit positive charge from *one specific point to another*, rather than moving it from *infinity* to some point. The work done in moving a unit charge, say, from point A to point B (see Fig. 13) is simply the *difference* in potential between points A and B. Thus, a *potential difference* of *1 volt* is said to exist between points A and B (in Fig. 13), if it requires an expenditure of energy of *1 joule* (10^7 ergs) to move

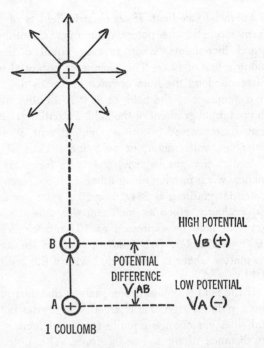

Fig. 13. The Potential Difference Between A and B is Equal to the Work Done in Moving a Unit (+) Charge from A to B

a positive charge of *1 coulomb* from point A to point B, along a line of force joining the two points. (Here point B is referred to as having a *higher* potential than point A.) Actually, it is not necessary that *we* perform work in moving the charge *against* the line of force. The *work may be done by the charge* in being *repelled* from point B to point A along a line of force. In either case, the potential difference is 1 volt if 1 joule of work is performed or recovered in moving 1 coulomb between points A and B. Note that the motion has to be with or against the field along a line of force. No work is performed in moving *across* a line of force, as we shall see presently.

Evidently, if the work performed in moving a unit charge from one point to another is equal to the potential difference (V) between the points, the work (W) done in moving any charge Q between the points must equal the *product* of the charge and the potential difference, or expressed as a formula,

$$W = QV$$

where the work (W) will be in *joules*, if the charge (Q) is expressed in *coulombs* and the potential difference (V) is in *volts*.

EXAMPLE 5: What work is done, when a charge of 25 coulombs is transferred between two points having a potential difference of 30 volts?

Solution: The work performed $W = QV = 25 \times 30 = 750$ joules. (1 joule $= 10^7$ ergs.)

Potential Gradient. If an electric field is of uniform strength, the potential changes smoothly in equal increments, when moving equal distances along a line of force. This change of potential with distance along the lines of force measures the relative steepness of the field, or as it is usually called, the **potential gradient** of the field. Potential gradient may be expressed in any convenient unit, such as volts/cm, volts/meter, or volts/inch. Thus, if the potential changes by, say, 254 volts for every 10 inches, when moving along a line of force, then the potential gradient is 254/10 or 25.4 volts per inch. Equivalently, since an inch contains 2.54 cm, the gradient may be expressed as 10 volts/cm, or as 1000 volts/meter. This gradient remains the same, no matter where it is measured within the uniform field.

If the field is **non-uniform**, such as that surrounding a point charge or a charged sphere, the potential does *not* change equally for equal increments in distance along a line of force. The potential gradient in such a non-uniform field, consequently, *changes from point to point* and it must be specified for a particular point within the field. The potential gradient is generally greatest near abruptly changing projections or sharply pointed areas on the surface of charged bodies, and it is these pointed areas that are most likely to "leak" electricity in the form of corona or brush discharge.

Equipotential Lines and Surfaces. Just as the surveyor draws contour lines on a map to specify all places of equal elevation or level, lines may be drawn through the electric field surrounding a charged body along which the **potential everywhere will be the same.** Such lines are called **equipotential lines.** Let us draw a few of these equipotential lines around a concentrated charge. (See Fig. 14.) From our previous formula for the potential in the vicinity of a point charge ($V = Q/k\,r$), we know that the potential is everywhere the same at a fixed distance "r" from such a charge. We therefore simply draw a series of circles of varying radii (r) around the point charge as center to obtain the equipotential lines in such a field. Note that these concentric circles are everywhere perpendicular to the lines of force emanating from the point charge.

By definition, the work done in moving a unit charge between two points equals the potential difference between these points. Since the potential along an equipotential line is everywhere the same, there is no potential difference between any points on such a line, and hence no work is done in moving

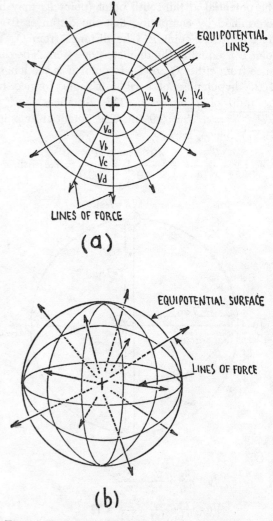

Fig. 14. Equipotential Lines (a) and Surface (b) around a Concentrated Charge

a charge along an equipotential line. We now understand why it is necessary that a charge move along a line of force so that work may be done by it or on it. Moreover, an electric current—which consists of charges in motion—will always flow along lines of force that exhibit a **difference in potential** and *never* along equipotential lines. This is the only way a current can perform useful work.

The concept of equipotential lines is easily extended to three dimensions. The potential near a point charge or charged sphere is, of course, everywhere the same on a **spherical shell** of a certain radius with the charge at the center (Fig. 14b). Such a shell is called an **equipotential surface.** The equipotential surfaces around a point charge or charged sphere consist of a series of concentric spherical shells.

STORING ELECTRICAL CHARGE— THE CAPACITOR

As we have seen, the potential at any point in space near a charged conductor is proportional to the charge on it. It follows that the potential right *at the surface* of the conductor must also be proportional to the charge on it. Moreover, the potential at the surface of a charged conductor, such as a sphere or a plate, must *everywhere be the same*, since otherwise the charges would *move* under the influence of a potential difference. We can express the proportionality between the charge and the potential anywhere on the surface of a charged conductor by the relation

$$Q = C V$$

where C is a proportionality constant known as the electrical **capacitance** of the conductor. Numerically, the capacitance is equal to the **charge required to bring a unit increase in potential.** For a **sphere,** as an example, the capacitance (in air) turns out to be **equal to its radius** in centimeters, if both the charge and the potential are expressed in electrostatic units (esu).

The Plate Capacitor. The capacitance of isolated conductors is not of great interest, since it is too small to permit storing a sizeable amount of charge. It is found, however, that the electrical capacitance of a conductor can be increased substantially by bringing a second conductor close to it. Such an arrangement of two conductors, separated by an insulator or **dielectric** (air, paper, etc.), is called a **capacitor** because of its ability to store electric charge. (The obsolete term "condenser" is still used occasionally.) Let us look at the action of a typical capacitor, consisting of two parallel conducting plates separated by an air dielectric. (See Fig. 15.)

As shown in the illustration, the two conducting plates have been connected by means of wires to the positive and negative terminals, respectively, of a battery, which serves as a source of electric charges (electrons). When the connection is first made, electrons rush out of the negative terminal of the battery and flow through the wire into plate *B*.

This plate, therefore, acquires an excess of electrons and becomes **negatively** charged. At the same time the influence of the **positive** battery terminal attracts electrons away from conducting plate A. An equal number of electrons will therefore flow **out of plate** A and through the connecting wire into the positive battery terminal. As a result, plate A acquires a **deficiency** of electrons and becomes **positively** charged. This initial rush of charges, when the capacitor is first connected to the battery, is known as the **displacement current** or the **charging current.** The current ceases when each of the plates has been charged to the *same* potential as that of its respective battery terminal, since then no *difference* of potential remains to provide the motive power for charges to flow. With each of the plates charged to the same potential as the respective battery terminal, the *difference* of potential (V) between the plates is, of course, exactly the same as that between the battery terminals. A capacitor, thus, always **charges itself to the voltage of the source.**

Each plate by itself, in the absence of the other plate, acquires, of course, a certain charge that depends on the capacity of the plate and the potential of the source $(Q = C V)$. As we have seen, the amount of this charge is small because the capacitance of an individual plate is low. When the two charged plates are brought close together, however, they establish a strong electric field between them, with lines of force extending from the positive to the negative plate. As indicated by the evenly spaced, straight lines of force in Fig. 15, the field between the plates is *uniform* in strength, except near the edges where the lines "fringe" out. Because of the presence of the field the opposite charges on the plates are strongly attracted toward each other and are concentrated on each plate. As a result more "room" becomes available on each plate for an inflow of *additional* charges. The capacitance of the combination, which is the **ratio of total charge stored to the potential difference** between the plates $(C = Q/V)$, is thereby **increased.**

A more sophisticated and accurate way of looking at the action of a capacitor is to consider the **potential** of the plates. Assume that the plates are initially separated by a sufficient distance to have no influence on each other. When connected to the battery, each plate will charge to the potential of the respective battery terminal, as we have explained before. The charge on each plate is determined by its capacitance and the potential of the battery terminal. Now let the negative plate (*B*) be

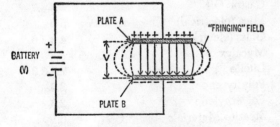

Fig. 15. Principle of Parallel Plate Capacitor

brought in close proximity to the positively charged plate (*A*). Since plate *B* is negatively charged, it will establish a **negative potential** in space at plate *A*. This negative potential will **subtract** from the positive potential of plate *A* and, hence, momentarily **lower** it. Since capacitance is the **ratio** of charge to potential ($C = Q/V$) and the charge has not yet changed, the **capacitance must have increased** with the lowered potential. With the capacitance increased, more charge will now flow into the plates, until the potential difference between the plates is again equal to that of the battery and a new balance is attained. This total charge will *remain* on the plates, even after they are disconnected from the battery.

Capacitance of Parallel-Plate Capacitor. By definition, the capacitance of *any* capacitor is the **ratio of the charge stored by its plates to the potential difference between them**, or expressed as a formula

$$C = \frac{Q}{V}$$

where the capacitance (*C*) will be in **farads**, if the charge (*Q*) is expressed in **coulombs** and the potential difference (*V*) in **volts**. The farad, however, is a very large unit and, hence, two smaller units are used in practice: one is the **microfarad** (abbreviated μf), which is a *millionth* of a farad ($1 \mu f = 10^{-6}$ farad); the other is the **micromicrofarad** (abbreviated $\mu\mu f$), which is a millionth of a microfarad or a *trillionth* of a farad (i.e., $1 \mu\mu f = 10^{-6} \mu f = 10^{-12}$ farad).

EXAMPLE 6: What is the capacitance of a capacitor that has a charge of 1/10 coulomb stored on its plates and a potential difference of 1000 volts between them?

Solution: $C = Q/V = 0.1/1000 = 0.0001$ farad $= 100 \mu f$.

In the case of a parallel-plate capacitor, the capacitance is easily computed. As shown in Fig. 16, the capacitance of this type increases directly with the *area* of either plate and with the dielectric constant (*k*) of the medium between them. The capacitance also *increases* as the separation (*d*) between the plates is made *smaller*, as we have seen before. Inserting the proper proportionality factor and taking into account the units used, a simple approximate formula for the capacitance of such a two-plate capacitor, turns out:

$$C\,(\mu\mu f) = \frac{k\,A}{4.45\,d}$$

where C = capacitance in **micromicrofarads** (1 farad $= 10^{12} \mu\mu f$)

A = area of one plate in **square inches**
d = separation between the plates **in inches**
k = dielectric constant of the medium

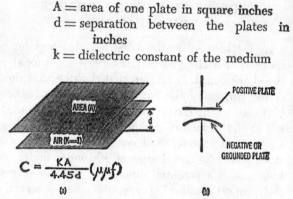

$$C = \frac{KA}{4.45d}\,(\mu\mu f)$$

(a)   (b)

Fig. 16. Capacitance of Parallel-Plate Capacitor (a) and Schematic Circuit Symbol (b)

The illustration (Fig. 16*b*) also shows the circuit symbol of *any* type of capacitor, used for schematic circuit diagrams. Note that the straight line indicates the positive or **high-potential** plate of the capacitor, while the curved line indicates the negative or low-potential plate of the capacitor. This plate is usually connected to a common circuit **ground**.

As is evident from the formula, capacitance also depends on the dielectric constant (*k*) of the medium. Although the dielectric constant of air is only 1 (approximately), it is frequently used as a dielectric in capacitors because it does not lose any of the charge supplied to the capacitor, while other dielectrics *do* waste a certain amount of it. Because of the low dielectric constant, air capacitors require large plates, and moreover, the plates cannot be spaced too closely because of the possibility of breakdown of the dielectric and **arcing over** between the plates, when their potential difference is high. Other dielectrics not only have a greater dielectric constant (*k*) but generally also a substantially greater **dielectric strength** against arc-overs. Table III lists the dielectric constants of some commonly used capacitor dielectrics.

TABLE III

Dielectric Material	Dielectric Constant (*k*)
Air	1.00
Paper	2.0-2.6
Castor Oil	4.3-4.7
Mica (electrical)	5-9
Glass (electrical)	4.2-7.0
Mycalex	8
Lucite	2.5-3.0
Polystyrene	2.6
Polyethylene	2.3
Steatite Materials	6.1

EXAMPLE 7: What is the capacitance of a paral-

el-plate capacitor having 2 in. x 2 in. plates, separated by 10-mil thick polystyrene dielectric? (1 in.
= 1000 mils)

Solution:

$$C = \frac{k\,A}{4.45\,d} = \frac{2.6 \times (2 \times 2)}{4.45 \times 0.01} = 234 \text{ micromicrofarads.}$$

Practice Exercise No. 2

1. Why does your hair "stand on end" when it is
vigorously combed on a dry day?

2. Explain what happens if the leaves of a charged
electroscope first converge and then diverge again,
when a test charge is gradually moved toward its metal
disk.

3. Why are two suspended pith balls first attracted
when a charged rod is brought near them, and then
repelled after contact? Why do the balls cling together
if one is brought in contact with a charged hard-rubber
rod and the other with a charged glass rod?

4. A copper sphere is mounted on an insulated stand.
Explain how you would charge the sphere *positively*
(a) by contact and (b) by induction.

5. How would you prove that no charge can reside
on the *inside* of a conducting body.

6. If 12 positive test charges experience a force of
360 dynes, when inserted into an electric field, what is
the strength (intensity) of the field?

7. Two small charged bodies with charges of +10
esu and −50 esu, respectively, are placed in air 10 cm
apart. What is the force between them?

8. Two concentrated equal charges in air repel each
other with a force of 1600 dynes over a distance of 30
cm. (a) What is the amount of each charge and (b)
what is the field intensity at the location of each
charge?

9. Equal charges of +20 esu each are placed at *two*
corners of an equilateral triangle having sides of 5 cm.
(a) Find the potential at the third corner and (b) that
at the center of the side joining the two charges. (Assume K = 1)

10. What work is done when 50 coulombs are transferred between two points having a potential difference of 120 volts?

11. The potential in a *uniform* electric field is found
to change by 12 volts every 3 inches. What is the potential gradient in volts/in and in volts/ft?

12. The charge on a parallel-plate air capacitor is
found to be 0.12 coulombs for a potential difference of
2400 volts between the plates. (a) What is its capacitance? (b) If the plates are separated by 1.15 mils,
what are the dimensions of each plate?

SUMMARY

Any two dissimilar materials may be electrically
charged by friction.

A body is charged when it has either a deficiency
or an excess of electrons. A **negatively** charged body
has an excess of electrons; a **positively** charged
body has a **deficiency** of electrons. Like charges of
electricity **repel** each other; unlike charges **attract**
each other.

All charges reside on the **outside** of a conducting
surface.

A charged body is surrounded by an **electric field.**
The **direction** of the field at any point is the direction in which a **positive** test charge would be urged
to move; the **intensity** (strength) of the field is the
magnitude of the force per unit charge. Both quantities are represented by the direction and number
(per unit area) of the **lines of force** of the field.

The force between concentrated point charges
varies **directly** with the **product** of the two charges
and **inversely** with the **square of the distance** between them. Coulomb's Law: $F = \dfrac{q_1\,q_2}{k\,r^2}$

A charge of 1 esu in free space placed 1 cm dis-
tant from an equal charge will repel the latter with
a force of 1 dyne. One coulomb equals 3×10^9 esu.

The **potential** at a point in an electric field is the
work done on or by a unit charge in moving from
infinity to the point. The potential (volts) expresses
the work (joules) per unit charge (coulomb) transferred.

A **potential difference** of 1 volt exists between two
points, if 1 joule of energy is expended to move a
charge of 1 coulomb between the points. **Potential
gradient** is the **change** of potential per unit distance
in an electric field.

The potential is everywhere the same along
equipotential lines and **surfaces,** and no work is
done in moving a charge along these lines or surfaces. Lines of force are **perpendicular** to the equipotential lines or surfaces.

The **capacitance** of a conductor (or capacitor) is
the **ratio** of the **charge** stored by it to its **potential**
(or potential **difference** between its plates): $C =$
Q/V. If the charge is expressed in **coulombs** and
the potential (or potential difference) in **volts,** the
capacitance is in **farads.** (1 farad = 10^6 μf = 10^{12}
$\mu\mu$f.)

The capacitance of a parallel-plate capacitor
(two plates) in micromicrofarads is: $C = \dfrac{k\,A}{4.45\,d}$,

where k is the dielectric constant of the material between the plates, A is the area of *one* plate (in
square inches) and d is the separation between the
plates (in inches).

CHAPTER THREE

MAGNETISM

Magnetism has been a familiar experience as long as electricity, but it took over two thousand years to discover the connection between the two. The ancient Greeks are said to have observed that pieces of a black mineral ore, known as *lodestone* or *magnetite*, were able to pick up small bits of iron. The Chinese discovered independently that splinters of lodestone rocks would orient themselves in the north-south direction, if freely suspended by a thread. These are among the fundamental properties of all magnetic substances. Until modern times all magnetic effects were studied by means of these weak natural magnets, since no others were available. After HANS CHRISTIAN OERSTED (1777-1851) discovered the relation between electricity and magnetism (in 1820), it became possible to make powerful artificial magnets by electrical means. These may exhibit magnetic properties either permanently or temporarily. All magnets used in practice are artificially produced.

FUNDAMENTAL MAGNETIC PROPERTIES

Magnets have the ability to attract iron. How strong this ability is depends on the material the magnet is made of. Artificial magnets are generally made of iron and steel, and are magnetized either by stroking with another artificial magnet or by being placed in the field of an **electromagnet** (described in a later chapter). The *harder* the steel the magnet is made of, the longer it will hold its magnetism, the property of permanence being called **retentivity**. Tungsten steel, chrome steel and cobalt steel make some of the most retentive permanent magnets. A number of materials, such as cobalt and nickel, have magnetic properties similar to iron and steel, but weaker, and are called **ferromagnetic** substances. Many other substances, known as **paramagnetic**, exhibit magnetism to a very slight degree, but not enough to be useful. A few substances are actually slightly *repelled* by a magnet and these are known as **diamagnetic**. It is an interesting fact that some of the ferromagnetic substances, though magnetically weak by themselves, make the hardest and most permanent magnets known when used as **alloys** with iron. Thus, the powerful modern **alnico** magnets consist of varying proportions of aluminum, nickel, cobalt, iron and copper. A new **cobalt-platinum** alloy, using no iron at all, is claimed to be 24 times stronger than even these powerful alnico magnets. In contrast, the *temporary* magnets we have mentioned are made of **soft iron** rods, that are contained in the coils of **electromagnets**. These soft-iron magnets can be powerfully magnetized, but they *retain their magnetism only while the electrical current is on*, except for a small amount, called **residual magnetism**.

It appears that magnetism is not distributed uniformly over the surface of a magnet, but is concentrated *near the ends*, in regions known as **poles**. A simple experiment will confirm this.

EXPERIMENT 4: Obtain a straight bar magnet, as shown in Fig. 17. You can make a bar magnet, if you have a horseshoe magnet available, by stroking a rod or bar of steel with one end of the horseshoe magnet a number of times. The steel bar will then be magnetized.

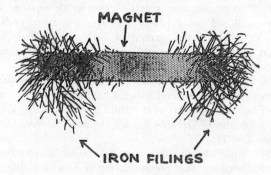

Fig. 17. Magnetic Attraction is Confined to the Ends (Poles) of a Magnet

If you now dip the bar magnet into a pile of iron filings, you will observe large clusters clinging to the bar near its ends, but practically no filings will be picked up near the center of the bar. This confirms that magnetism is confined principally to the ends (poles) of a magnet. Moreover, accurate tests show that the two poles of a magnet have *exactly the same strength*.

EXPERIMENT 5: Suspend a bar magnet from a thread near its center, so that it may freely turn about its axis. (See Fig. 18.) After a few oscillations the magnet will point in a general north-south direction, as you can easily check with an inexpensive **magnetic compass**, based on the same principle (Fig. 18). No matter how often you repeat the experiment, you will note that the *same end* of the magnet always points in the northerly direction, while the other end always points (approximately) south. For this reason, the end of the magnet that always points toward the northern regions of the earth is called the **north-seeking or north (N) pole**, while the other end is termed the **south-seeking or south (S) pole**. This is the principle of the magnetic compass, which has been of greatest importance to navigation.

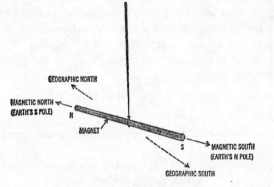

Fig. 18. A Suspended Magnet Always Rotates to the North-South Direction

The reason that a magnet or compass needle will always turn to the north-south direction is that the earth itself is a giant magnet with two poles. The magnetic poles do *not* coincide, however, with the earth's geographic poles, magnetic north being located in northern Canada, almost 1500 miles from geographic north. Magnetic south is at the opposite point of the globe, at the same distance from geographic south. Moreover, since like poles repel and unlike poles attract (as we shall presently see), the north-seeking pole of a compass actually points toward the **magnetic south pole**, while the south-seeking pole points toward the **magnetic north pole** of the earth. To avoid confusion, **the south magnetic pole is designated as magnetic north and the north magnetic pole as magnetic south.** Since there is still a considerable discrepancy between magnetic and geographic north, corrections for this error, known as **declination**, must be made at each specific location.

Either pole of a magnet can be used to attract a

piece of iron, but the poles of two magnets will not always attract each other. An Italian soldier, Peregrinus, discovered as early as 1269 A.D. that the north pole of a magnet will *repel* the north pole of another magnet; he found this to be true also for the south poles of two magnets. When he approached the *north* pole of one magnet with the *south* pole of another magnet, however, he found that these would *attract* each other. You can easily verify this fact by approaching the north pole of a compass needle (usually painted blue) with the north pole of a bar magnet. You will find that the point of the needle is violently repelled and the needle will rotate until its south-seeking pointer is opposite the north pole of the magnet. If you now turn the bar magnet around so that its south pole faces the south point of the compass needle, the needle will again be repelled and whirl around to bring its north pole into alignment with the south pole of the bar magnet. (See Fig. 19)

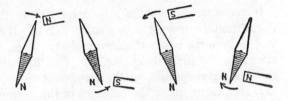

Fig. 19. Like Poles Repel. Unlike Poles Attract

From this behavior we deduce the general rule that **unlike magnetic poles attract each other and like poles repel.** This is similar to the rule in electrostatics, that unlike charges attract each other and like charges repel, except that the poles of a magnet always occur in *pairs* and *cannot be isolated* like electric charges. As a matter of fact, *poles are not essential* to magnetic behavior. If you bend the poles of a horseshoe magnet together or join the ends of a magnetized ring, you will have a perfectly good magnet that does not exhibit any poles. If you break the ring into two parts, however, each of the pieces will again exhibit north and south poles at its ends. Moreover, you can break each of the pieces into as many parts as you wish and, again, each of the broken bits will show a north pole and a south pole near its two ends. This behavior indicates that magnetism is associated with the molecular and atomic structure of matter, as we shall see later on.

Paralleling the **induction** of charges in electrostatics without contact, there is the similar phenomenon of **induced magnetism,** as demonstrated by the following experiment:

EXPERIMENT 6: Place one pole (north or south) of a strong steel magnet at the edge of a table and attach several tacks or small nails to the pole in chain fashion, as illustrated in Fig. 20. You will find that at least four or five *unmagnetized* tacks will cling to the magnet and to each other without difficulty.

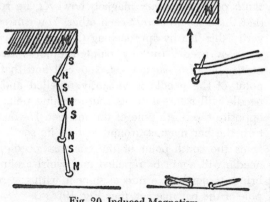

Fig. 20. Induced Magnetism

If you now remove the uppermost nail in the chain from the magnet with a small pair of pliers, you will see the whole chain crumble to pieces, as shown in the illustration (Fig. 20). This indicates that the magnetization of the soft-iron tacks was only *temporary*, under the influence of the magnet. Since none but the uppermost tack were in direct contact with the magnet, the temporary magnetization of the soft-iron bits is known as **induced magnetism**. It can also be shown that the pole of the magnet induces an opposite pole in the nearest tack (north induces south, as shown) and each of the tacks, in turn, induces an unlike pole in the end of the object clinging to it. If it were otherwise, the nails would repel rather than attract each other.

All the properties of magnetization we have discussed are based on certain **alignments** of large numbers of atoms (called **domains**), according to the theory to be described later on. These domain alignments are disturbed if a magnet is jarred, hammered, or heated, and partial demagnetization takes place. If a magnet is heated to a certain critical temperature, called the **Curie point**, the magnetic alignment is completely upset by the thermal vibrations of the molecules, and all magnetism disappears. This, then, is one way of demagnetizing a permanent magnet. A better and faster way is to place the magnet in the field of an alternating current, in a demagnetizer. This method is based on electromagnetic properties, which we shall describe in a later chapter.

LAW OF FORCE BETWEEN MAGNETIC POLES

The French physicist Charles A. Coulomb, who discovered the law of force between electric charges, also experimented with magnetic poles, using the same torsion balance method as for electrostatic forces. By suspending a long bar magnet from a wire and bringing the pole of another magnet near a pole of the bar magnet, the poles either attracted or repelled each other, and in the process, twisted the wire. A series of painstaking experiments led him to conclude (in 1785) that the **law of inverse squares** was also valid for the force between point (isolated) poles. (Poles may be considered "isolated" from each other at the ends of a long, thin magnet.) Specifically, Coulomb's Law for magnetic poles states that the **force between point poles is directly proportional to the product of the pole strengths and inversely proportional to the square of the distance between the poles.** This relation is expressed mathematically in the following form:

$$F = \frac{m_1 \times m_2}{\mu \, r^2}$$

where *F* is the force (of attraction or repulsion) in *dynes*

m_1 and m_2 are the strengths of the two poles, respectively

r is the distance between the poles in *centimeters*

and μ depends on the medium in which the poles are located.

The constant μ, called the **permeability** of the medium, shows the relative ease of magnetization of a material. Its value is unity for a vacuum and practically that for air and other gases. As we shall see later on, the permeability is very high, in the order of several thousand, in iron and ferromagnetic materials.

Coulomb's law of force serves to define the **unit pole** of magnetic strength: a unit pole is of such strength (m_1) that it will exert a force of one dyne upon an equal pole (m_2) in vacuum, when placed at a distance of 1 cm away from it. This unit pole is sometimes referred to as the **electromagnetic unit** (e.m.u.) of pole strength.

EXAMPLE 1: Find the force of repulsion between an isolated north pole (of a long, thin magnet) of 40 e.m.u. strength, which is placed in air at a distance of 6 cm from a like pole of 30 e.m.u. strength. *Solution:*

$$F = \frac{m_1 m_2}{\mu r^2} = \frac{40 \times 30}{1 \times 6^2} = \frac{1200}{36} = 33.3 \text{ dynes}$$

EXAMPLE 2: Compute the *net force* between two 30-cm long bar magnets that are placed in line on a table, with their south poles 10 cm apart and their north poles 70 cm apart. Assume that the strength of each pole is 140 e.m.u. and is concentrated at a point at the ends of each magnet. The medium is air. (Fig. 21.)

Solution: (See Fig. 21.) Four forces act on the magnets, two of repulsion between the S poles and between the N poles, and two of attraction between N and S poles. Thus, the forces of **repulsion** are:

$$\frac{140 \times 140}{1 \times 10^2} = \frac{19,600}{100} = 196 \text{ dynes};$$

$$\frac{140 \times 140}{1 \times 70^2} = \frac{19,600}{4,900} = 4 \text{ dynes}$$

The forces of **attraction** are:

$$\frac{140 \times 140}{1 \times 40^2} = \frac{19,600}{1,600} = 12.25 \text{ dynes};$$

$$\frac{140 \times 140}{1 \times 40^2} = 12.25 \text{ dynes}$$

Hence, the **net force** will be $(196 + 4) - (12.25 + 12.25) = 175.5$ **dynes repulsion.**

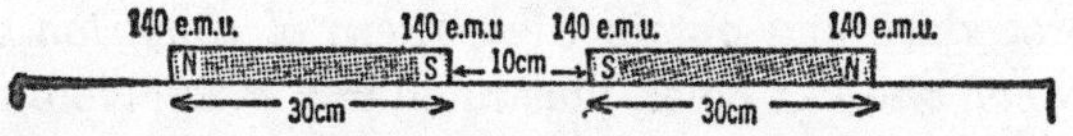

Fig. 21. Force Between Two Bar Magnets (Example 2)

MAGNETIC FIELDS AND LINES OF FORCE

A permanent magnet exerts a force on a piece of iron or on another magnet placed at some distance from it. We like to explain such mysterious "action at a distance" by a **field of force** which extends over the space where the effects of the force can be felt. To explain the pull of the earth on an object (weight), we speak of a gravitational field and the region of influence around electric charges is referred to as an electrostatic field. Similarly, the region surrounding a magnet, where its influence can be detected, is known as the **magnetic field of force.** Throughout this region magnetic poles or substances will be subjected to a force that varies in direction and amount as the pole or substance is moved about in the field. The direction in which a free (isolated) **unit north pole** would be urged to move defines the **direction of the magnetic field** at that point. The fact that such an isolated north pole

actually cannot exist, does not detract from the convenience of the concept. In practice, a small compass needle approximates the action of such a **test pole.**

The strength or **intensity of the magnetic field** at a point is defined as the **force** that would be exerted on a **unit north pole** placed at that point. The unit of field intensity is called the **oersted,** after the Danish physicist HANS CHRISTIAN OERSTED (1777-1851), who discovered the fundamental electromagnetic actions. The oersted, thus, represents the intensity of a magnetic field in which a **unit magnetic pole experiences a force of one dyne.** Accordingly, if a pole of strength m experiences a force of F dynes at a point in a magnetic field, the field intensity (symbol H) at that point is

$$H = \frac{F}{m} \text{ (oersteds)}$$

We can easily derive the field intensity around a pole of known strength M by substituting Coulomb's law for the force (F) in the formula above. Thus,

$$H = \frac{F}{m} = \frac{\dfrac{M \times m}{\mu r^2}}{m} = \frac{M}{\mu r^2}$$

where r is the distance from the pole and μ is the permeability.

EXAMPLE 3: An isolated north pole of 30 e.m.u. strength is placed (in air) in the magnetic field surrounding a magnetic pole of unknown strength and is repelled by it with a force of 240 dynes. If the distance between the north pole and the unknown pole is 5 cm, what is (a) the field intensity at that point and (b) the strength and polarity (N or S) of the unknown pole?

Solution:

(a) field intensity $H = \dfrac{F}{m} = \dfrac{240}{30} = 8$ oersteds.

(b) strength of unknown pole $M = H \mu r^2 = 8 \times 1 \times 5^2 = 200$ e.m.u. Since the force is one of repulsion, the unknown pole must be a north pole.

Lines of Force. As was the case for electrostatics, **lines of force** may be drawn to represent the configuration of a magnetic field. The *direction* of the magnetic field at a point may be shown by drawing the lines of force in the direction in which a unit north pole would be urged to move.

The *strength* of the field may be shown by drawing a certain number of lines per unit area. For a bar magnet, for example, the direction of the lines of force could be ascertained by carrying a small compass needle (serving as test pole) around the magnet, as shown in Fig. 22a. The needle will automatically set itself *parallel* to the lines of force at any point, thus indicating the direction of the sum-total of all the forces (resultant) upon it. Fig. 22b indicates the portrayal of the lines of force about a bar magnet that could be obtained in this way.

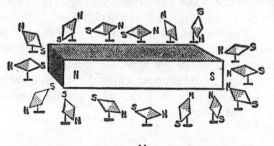

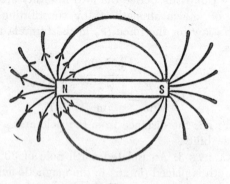

(b)

Fig. 22. Exploring the Field About a Bar Magnet with a Compass Needle (a) and Representation of the Field by Lines of Force (b)

Note that the lines issue from the north pole and terminate on the south pole, since this is the direction in which a unit north pole would tend to move.

A much better way to investigate the configuration of a magnetic field is to place a thin sheet of cardboard over a magnet, sprinkle some fine iron filings over it, and tap it gently. Each individual filing then becomes a **temporary magnet by induction** and aligns itself, as the compass needle, in the direction of the lines of force of the field at the particular point. Let us see how this works out for a number of typical fields.

EXPERIMENT 7: Place a sheet of cardboard over a bar magnet, sprinkle iron filings on it and tap lightly until the pattern shown in Fig. 23 emerges. Note that lines of force never cross each other and that they seem to repel each other sideways.

Now place the bar magnet *vertically* on one end put the cardboard and iron filings on top of the other pole, while supporting the magnet with your hand. Repeat the previous procedure, tapping gently, until the filings arrange themselves in the pattern of an isolated (north or south) pole shown in Fig. 24. (The pole may be considered *isolated*, since the other pole of the magnet has little effect upon it in this position.) Repeating the procedure for the other pole of the vertical magnet, the same general pattern will be seen to emerge.

EXPERIMENT 8: Obtain another bar magnet similar in size and strength to the one used for experiment 7. Place *both* magnets *vertically* on a table or in a (vise) support, a few inches apart, so that a free north pole and a free south pole extend vertically up. Place a cardboard on top of the two free poles, sprinkle iron filings on it, and tap. The resulting pattern will show lines of force extending from the north pole to the south pole, indicating *attraction* between the two. (See Fig. 25.) Note the crowding together of the lines near the poles, where the field is strong, and their spreading apart *between* the poles, where the field is weaker.

Now *reverse* the position of *one* of the bar magnets, so that either two north poles or two south poles face vertically upward. Repeat the previous procedure and obtain the pattern of *repulsion* between two like poles, illustrated in Fig. 26. Note the general similarity of the field patterns for unlike and like poles in Figs. 25 and 26, respectively, to those illustrated in Fig. 10 for *electrostatic charges*.

EXPERIMENT 9: Obtain an inexpensive horseshoe magnet, place it flat on a table, and put a cardboard over it. Sprinkle with iron filings and tap, to obtain the field pattern shown in Fig. 27.

Now hold the magnet in *vertical* position or place it in a vise so that the poles extend vertically upward. Place a cardboard on top of the poles and repeat the previous procedure, while holding the paper with your hand. The resulting field pattern for a vertical horseshoe magnet is shown in Fig. 28.

Experiments 7 through 9 demonstrate graphically the usefulness of the lines-of-force concept for mapping magnetic fields. When Michael Faraday first introduced the concept, he thought of the lines of force as having physical reality and visualized them as "elastic tubes" or rubber bands. Thus, the attraction between two unlike poles, for example, could be explained mechanically by the tension of these elastic tubes: in attempting to shorten them-

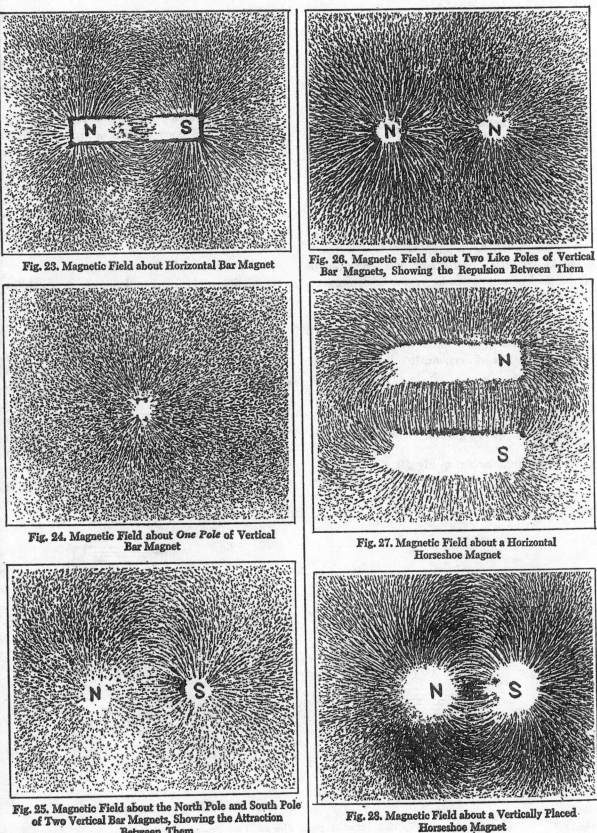

Fig. 23. Magnetic Field about Horizontal Bar Magnet

Fig. 26. Magnetic Field about Two Like Poles of Vertical Bar Magnets, Showing the Repulsion Between Them

Fig. 24. Magnetic Field about *One Pole* of Vertical Bar Magnet

Fig. 27. Magnetic Field about a Horizontal Horseshoe Magnet

Fig. 25. Magnetic Field about the North Pole and South Pole of Two Vertical Bar Magnets, Showing the Attraction Between Them

Fig. 28. Magnetic Field about a Vertically Placed Horseshoe Magnet

selves the lines of force tend to pull the poles together.

Modern theory considers the lines of force as an *imaginary*, but highly useful concept for mapping magnetic fields and calculating their effects. To make quantitative determinations appropriate units must be assigned. A **single line of force** represents the unit of **magnetic flux** in a field and is called the **maxwell**. The **total magnetic flux** (symbolized by the Greek letter phi, ϕ) is a magnetic field, consequently, is measured by the **total number of lines of force**, or maxwells. (By convention, the magnetic flux is said to issue from the N pole of a magnet and travels to the S pole.) Since the maxwell represents a very small quantity of flux, a larger unit, called the **weber**, is frequently employed. One weber is equal to 100,000,000 or 10^8 *maxwells*. The strength of the field in any particular region is determined by the number of lines of force traversing a unit area; that is, by the *flux per unit area* (ϕ /A) in that region. This quantity is called the *flux density* (symbol B) and it is expressed either in gauss (*maxwells/cm²*) or *webers/m²*. It follows from this that 1 weber/m² equals 10^4 *gauss* (that is, $\dfrac{1 \text{ weber}}{1 \text{ m}^2}$

$= \dfrac{10^8 \text{ maxwells}}{10^4 \text{ cm}^2} = 10^4$ gauss). These relations are conveniently summarized in Fig. 29.

Since flux density is *flux per unit area*, it follows that the total flux (ϕ) through a region in a magnetic field may be obtained by multiplying the flux density (B) by the area (A) of the region. Expressed mathematically, the total flux

$$\phi = B \times A$$

EXAMPLE 4: A flux density of 20,000 gauss is observed in a circular gap of 2 cm radius. What is the total flux through the gap?

Solution: The area of the gap $= \pi r^2 = 3.14 \times 2^2 = 12.58$ sq.cm. Hence the flux $\phi = B A = 20,000 \times 12.58 = 252,000$ maxwells (lines of force) or 0.00252 weber.

The flux density is the flux per unit area *induced* in a certain substance. It depends, therefore, on the field intensity as well as on the **permeability** of the medium. The simple relation between flux density (B) and field intensity (H) is

$$B = \mu \times H$$

This relation is also used to define the **permeability of a medium** as the **ratio of flux density to field intensity**; that is,

$$\mu = B/H$$

In vacuum or in air the permeability is unity (approximately) and, hence, the flux density (in gauss) is *numerically equal* to the field intensity (in oersteds) for these media. We shall see in the chapter on electromagnetism that the permeability of magnetic substances is several thousand times that of air, permitting values of flux density in these substances that are far greater than can be achieved in air. Magnetic materials (iron and ferromagnetic alloys) are used, for this reason, to guide the flux in **magnetic circuits**.

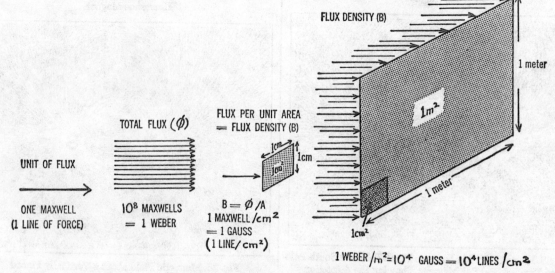

UNIT OF FLUX

ONE MAXWELL
(1 LINE OF FORCE)

TOTAL FLUX (ϕ)

10^8 MAXWELLS
= 1 WEBER

FLUX PER UNIT AREA
= FLUX DENSITY (B)

$B = \phi /A$
1 MAXWELL /cm²
= 1 GAUSS
(1 LINE/ cm²)

FLUX DENSITY (B)

1 meter

1m²

1 meter

1cm²

1 WEBER /m² = 10^4 GAUSS = 10^4 LINES /cm²

Fig. 29. Units of Magnetic Flux and Flux Density

THEORY OF MAGNETISM

Modern research has established that magnetism is *not* a fundamental phenomenon, but only one aspect of *electrical* behavior. The Danish physicist Oersted discovered in 1820 that an **electric current** in a conductor is always surrounded by a magnetic field. This is the basis of *electromagnetism*. But it was not until recently that it became evident that **magnetism itself must be attributed to electrical charges in motion.** Specifically, the **motion of the** electrons within the atom constitutes an electric current and this tiny current exhibits a magnetic effect.

The orbital electrons within the atom not only revolve about the nucleus, but they also *spin about their own axis* and it is this motion that is responsible for magnetic effects. In ordinary, non-magnetic materials the same number of electrons spin clockwise about their axes as spin counterclockwise and, consequently, there is no *net* internal spin or motion of charge present. Since the electron spins cancel out in these materials, they cannot be magnetized. In the atoms of ferromagnetic materials, however, *more electrons spin in one direction than in the other*. These **uncompensated** or **unbalanced** electron spins create small magnetic "twists" (called **magnetic moments**), which make each atom of a ferromagnetic material a tiny magnet. Since these atomic magnets are oriented in *random* directions, they do not produce observable overall magnetic effects.

To explain large-scale magnetic behavior the *alignments* of the uncompensated electron spins must be considered. It is believed that the uncompensated spins in a magnetic material have the same direction (i.e., are parallel) throughout a tiny region of some 10^{15} atoms, called a *domain*. (The head of a common pin contains at least 6000 domains.) The parallel electron spins within a single domain produce an intense magnetic field in its vicinity. Unfortunately, the domains in a magnetic material, such as iron, are oriented in all conceivable (random) directions and, hence, the **internal fields of the domains ordinarily cancel out.** (See Fig. 30a.) But when the material is magnetized, by subjecting it to an *external* magnetic field, the domains begin to rotate—a few at a time—to *align themselves with the external field*. (Fig. 30b.) This is a gradual process that depends on the strength of the magnetizing field and the duration of magnetization. Eventually, when all the domains have "jumped" into alignment with the external field, complete magnetization is attained and **magnetic saturation** is said to occur. (See Fig. 30c.)

Practice Exercise No. 3

1. What would you do to identify the poles of an unmarked permanent magnet if you had (a) another marked magnet, (b) only a piece of string?

2. Why is repulsion a better test for polarity than attraction?

3. A steel bar is stroked from left to right with the

MAGNETIC FIELD

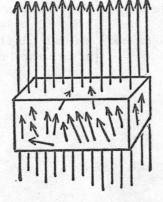

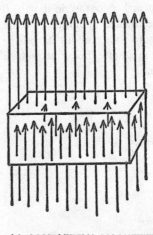

(a) UNMAGNETIZED

(b) PARTIALLY MAGNETIZED

(c) COMPLETELY MAGNETIZED (SATURATION)

Fig. 30. Alignment of Domains When a Material is Magnetized

south pole of a permanent magnet. What is the polarity of the right end and that of the left end after magnetization?

4. A bar of steel that has been magnetized by stroking with another magnet exhibits a north pole at *each* end. How is this possible?

5. Two equal poles 6 cm apart in air repel each other with a force of 64 dynes. What is the pole strength?

6. A 10-cm long bar magnet with pole strengths of 300 e.m.u.—each is placed in a straight line with another bar magnet that is 12 cm long and has pole strengths of 400 units each. If the north poles are 3 cm apart, what is the force between the magnets?

7. (a) Find the intensity of the magnetic field about a pole of 400 units strength at a distance of 10 cm. (b) If a north pole placed at that point is repelled with a force of 60 dynes, what is its strength?

8. The field intensity within a hollow rectangular frame of 12 cm by 12 cm is found to be 500 oersteds. What is (a) the flux density, and (b) the total flux passing through the frame?

SUMMARY

Magnets have the ability to **attract iron** and other **ferromagnetic** substances. Magnets may be either **natural** (lodestone, magnetite), or they can be made artificially by magnetization with another magnet or in an **electric field**.

Magnets made of **hard steel** and ferromagnetic materials (cobalt, nickel, aluminum, etc.) are **permanent**; **soft iron** can be magnetized only **temporarily**, but retains a small **residual magnetism**.

Magnetism is concentrated near the ends of a magnet in the **poles**. The **north-seeking** (N) pole points toward geographic north (approximately), the S pole toward geographic south, if the magnet is left free to rotate about its axis.

Unlike magnetic poles **attract; like** poles **repel** each other.

Magnetism may be **induced in a magnetic material** by bringing it into the field of a magnet. Induced magnetism in soft iron is only **temporary**.

When a magnet is hammered or heated, it loses its magnetism.

The force between point poles is **directly proportional to the product of the pole strengths and inversely** proportional to the square of the distance between them. ($F = m_1 m_2/\mu r^2$)

A **unit magnetic pole** (e.m.u.) placed at a distance of 1 cm from an equal pole in vacuum or in air repels it with a force of 1 dyne.

The **intensity** (strength) of a magnetic field at a point is the force in dynes exerted on a **unit north pole** at that point. It is measured in oersteds. ($H = F/m = M/\mu r^2$)

Lines of force are used to map the configuration of a magnetic field and calculate its effects. A single line of force (1 maxwell) is the unit of **magnetic flux**. One weber equals 10^8 lines of force or 10^8 maxwells. The **total magnetic flux** (ϕ) is equal to the total number of lines of force, or maxwells. It extends from the north to the south pole of a magnet.

Flux density is the flux per unit area ($B = \phi/A$). It is measured in gauss (maxwells/cm^2) or webers/m^2 ($= 10^4$ gauss). Total flux is the **product** of flux density and area. ($\phi = B \times A$.)

Permeability is the ratio of flux density to field intensity. ($\mu = B/H$.) It is a measure of the relative ease of magnetization.

Magnetism is caused by **uncompensated electron spins** in the atoms of ferromagnetic materials. The spins in a "domain" of about 10^{15} atoms have the same direction and produce intense, but random oriented magnetic fields. Magnetization is produced by aligning the domains under the influence of an external magnetic field.

CHAPTER FOUR

SOURCES OF ELECTRIC CURRENT

An electric current consists of charges (electrons) in motion. To make the charges move, some driving or electromotive force (emf) must be provided, as we have seen in Chapter 1. Essentially, the electromotive force that makes free electrons flow through a conductor is their repulsion by negative charges and the equal attraction to positive charges. Hence, electron flow always takes place from a negatively charged source to a positively charged "sink." Most substances are electrically neutral, however, since the positive and negative charges within their atoms are in balance. A certain amount of work must therefore first be performed to separate or displace the electrical charges, so as to create a surplus of electrons (negative charge) in a source of electric current and a deficiency of electrons (positive charge) in the sink. But, as we have seen in Chapter 2, the work done in transferring charges from one point to another is the difference in potential between the points. Thus, the work performed in creating a difference of potential between a current source and sink provides the electromotive force that causes free electrons to flow. Of course, in flowing through a conductor, the electrons return the work, either by heating the conductor or doing useful tasks in a "load." Moreover, the flow of charges between source and sink equalizes the imbalance of charges and the current flow eventually will stop, unless the potential difference or emf is maintained in some way.

Since all matter is essentially electrical, a potential difference or emf that will serve as a source of electric current can be obtained in many different ways. We have already learned that rubbing two insulators together and then moving them apart (glass and silk, for example) will develop a potential difference between the two by the separation of charges. In addition to electricity obtained by mechanical friction, we have referred to other sources of electric current, including heat (thermoelectricity), light (photoelectricity), pressure (piezoelectricity), chemical action (electrochemistry), and mechanical motion of a conductor in a magnetic field (electromagnetism). These are by no means the only sources, and we could consider, for example, the direct radiation of charged particles from the nuclei of radioactive materials or electromagnetic radiations. The six sources listed above are the chief forms of primary energy capable of producing electricity. Of these the only significant source of commercial electrical power is the mechanical action of electric generators, which is based on electromagnetic principles we shall study in a later chapter. As a source of emergency and mobile electric power the chemical action produced in batteries is also of some importance and, hence, we shall devote a separate chapter to electricity from chemical action (Chapter Five). The other sources mentioned are at the present time considered too weak for electric power exploitation, but they have a great many applications in electrical instruments and in the field of electronics.

ELECTRICITY FROM HEAT (THERMAL EMF)

When two *dissimilar* metals, such as a copper and an iron wire, are joined together at both ends and one of the junctions is at a *higher temperature* than the other, a difference of potential (emf) appears between the wires and current flow through the wires will result. Current flow due to a thermal emf is known as the thermoelectric effect; it was discovered in 1821 by the German physicist THOMAS J. SEEBECK (1770-1831). You can make a simple experiment to verify the effect.

EXPERIMENT 10: Twist together the ends of a copper (bell) wire and an iron wire, cut the copper wire near its center and connect the free ends to an inexpensive galvanometer or microampere meter. (The meter will indicate *zero* current flow through the wires.) Now keep one of the junctions at a low temperature by inserting it in a glass of ice water and heat the other junction with a candle or bunsen burner, as shown in Fig. 31. As soon as this is done, the meter will deflect to indicate a current of several microamperes (1 microampere = one millionth ampere) flowing through the wires. The greater the difference in temperatures between the junctions, the greater will be the current, up to a certain point.

If the galvanometer is deflecting properly, the current must be flowing from the cold to the hot

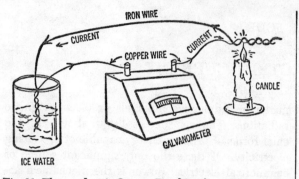

Fig. 31. Thermoelectric Current Resulting from Temperature Difference Between Junctions of Dissimilar Wires

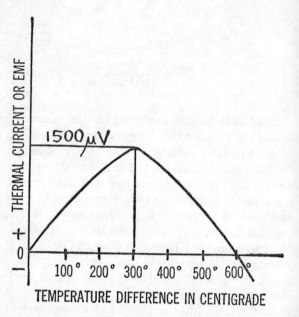

Fig. 32. Graph of Thermal Current (or Emf) against Temperature Difference for Copper-Iron Junctions

junction through the copper wire, and from the hot to the cold junction in the iron wire. You can verify this by reversing the hot and cold junctions, exchanging the positions of the ice water and the candle. The meter will now attempt to deflect in the opposite direction (but cannot, of course), indicating that the current has also reversed. To make it deflect properly, you will have to reverse the connections to the meter terminals.

In case you have available an accurate thermometer with a range of about 600° C, you can obtain some quantitative data on the thermoelectric effect. Start with approximately equal junction temperatures and slowly increase the temperature difference. (You can do this by starting with a candle on one end and a glass of *hot* water of equal temperature on the other. Let the water cool off and after it reaches room temperature drop ice cubes into it.) If you keep an accurate record of the temperature *difference* between the copper-iron junctions, you will be able to plot a graph of galvanometer current versus temperature difference, as shown in Fig. 32. This graph shows that the current goes up almost *linearly*, up to a temperature *difference* of 300° Centigrade (540° F), but *decreases* for temperature differences greater than that, reaching zero again for a temperature difference of about 600° C (1080° F). If you make the temperature difference still greater, the current will start to flow in the *opposite* direction.

Instead of current, usually the **emf** produced between the junctions is measured. This is very small, in the order of some *millionths* of a volt. For copper-iron junction, this emf is found to be about 7 *microvolts* (millionths of a volt) per degree centigrade of temperature difference between the junctions. For a *copper-constantan* junction, a pair of metals frequently used, the emf developed is as high as 40 *microvolts* per centigrade degree of

temperature difference. Other combinations have emf's somewhere in between.

Applications. You may have noted in experimenting with various metal junctions that some provide a sufficiently high thermal emf to make it unnecessary to artificially cool one junction. Heating one junction and leaving the other at room temperature is usually enough to provide a readable meter deflection. As a matter of fact, *one heated junction* of dissimilar wires is all that is required to show the effect. The free ends of the wires may be connected to a galvanometer, which then constitutes the *cold junction* and measures the thermal emf or current at the same time. Such a single junction of two different metals that are twisted, brazed or riveted together at one end, is called a thermocouple. A typical thermocouple for commercial ap-

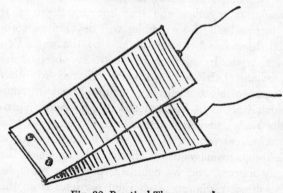

Fig. 33. Practical Thermocouple

plications is illustrated in Fig. 33. The free ends are connected to a measuring device.

Thermocouples are not used at the present time to furnish electric power, since the effect is small. It is a fairly simple matter, though, to connect a number of couples together (in series), in a so-called **thermopile, to multiply the emf** to a desired voltage. A few large thermopiles have been constructed, capable of producing currents of several amperes, which may eventually develop into practical **thermogenerators.**

The chief use of thermocouples is for measuring work. Thermocouples permit measuring a wide range of temperatures and are known as **pyrometers,** when so used. The thermocouple is inserted into a flame or furnace whose temperature is to be measured and the wire ends are connected to a sensitive meter, calibrated *directly* in *degrees of temperature*. Iron-constantan thermoelectric pyrometers permit temperature readings up to 1300° C and platinum/platinum-rhodium couples have a range up to 1600° C. (*Constantan* is an alloy containing 55% copper and 45% nickel.)

Thermocouples are also frequently used for measuring small **alternating** or **direct currents.** In this application the a-c or d-c current to be measured flows through a high-resistance **heater wire,** which heats the thermocouple junction. The free ends of the couple are connected to a sensitive *d-c* milliammeter, which measures the thermal current through the couple, and *indirectly*, the current through the heater. The instrument is very useful, since it can measure either d.c. or a.c. up to very high frequencies.

Thermoelectric Effect. The production of a thermal emf is a complex business and we can only hint at an explanation here. When two dissimilar metals are pressed together, free electrons pass haphazardly in both directions across the junction. Because of the different atomic structure of the metals, electrons pass more readily in one direction across the boundary than in the other. This results in a displacement of charges, making one metal positive and leaving the other negative. A difference of potential, known as **contact potential,** appears between the metals and this potential is influenced by the temperature of the junction. By keeping one junction of the two metal strips at a higher temperature than the other junction, the unequal drift of electrons past each junction maintains a difference in contact potentials, and a constant thermal electromotive force can be obtained.

ELECTRICITY FROM LIGHT
(PHOTOELECTRIC EMF)

As early as 1888 experiments showed that freshly polished zinc would lose a *negative* charge of electricity when exposed to ultraviolet light. It was found later that a large number of substances showed this **photoelectric effect,** when they were illuminated by ordinary, ultraviolet or infrared light. In the case of the polished zinc and similar light-sensitive materials, electrons are actually *emitted* from the surface of the material and **photoemission** is said to take place. Actually, there are three types of photoelectric devices, classified either as **photoemissive, photovoltaic,** or **photoconductive.** Between them these exciting gadgets account for a multiplicity of the control application we encounter every day. Photoelectric devices may be made to operate a relay whenever a beam of light falls upon the device. The relay, in turn, can open a garage door from a car's headlights, automatically dim headlights, operate a telegraph to permit transmission of messages by means of light beams, and so on. In contrast, the *interruption* of a beam of light, when an object passes between a **photocell** and a light source, can be made to *cut off* the current to a relay, which will then either open or close its contacts. This, in turn, can actuate a mechanical register for *counting* the objects passing the photocell, start an escalator, open a door, etc. All these applications and many more are controlled by electricity freed by light.

Photoemission (Phototubes). When light falls upon a **photoemissive** material, such as cesium, strontium, lithium, barium or other **alkali metals,** part of the light energy (called **photons**) is transferred to the free electrons within the material and literally kicks them out from the surface of the "emitter." To make the emitted electrons do useful work, they must be *collected* in some way and then made to flow through a conducting circuit that contains a "load." A **positively charged** metal plate, placed a short distance from the photoemitter, can serve as a collector by attracting the electrons. After arriving at the plate, the electrons are made to flow through an external conducting circuit. In practice, both the electron emitter (called **photocathode**) and the collector plate (**anode**) are enclosed in a sealed glass envelope, which is either highly **evacuated** or filled with an **inert** gas (argon, neon). Such an arrangement is called a **phototube.** The tubes must be evacuated to prevent the emitted electrons from being scattered in all directions by collisions with air molecules. The addition of an inert gas can be used to multiply the current passing through the tube. As the electrons collide with gas atoms, they knock out orbital electrons from the outer shell of the atom, thus electrifying or ionizing the gas. The ionized gas greatly increases the total current through the tube.

Fig. 34 shows a few typical phototubes. Note that the photocathode (emitter) is usually large in area to expose as much surface to the light as possible. It is generally in the form of a cylinder

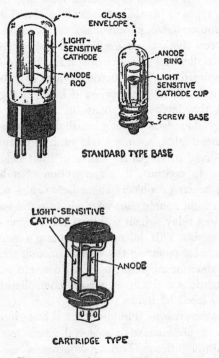

Fig. 34. Typical Forms of Phototubes

that surrounds a metal rod, which serves as **anode** or collector of electrons. A *positive* voltage must be applied to this anode to operate the tube. When this is done, the amount of current flowing through the tube and the external circuit is directly proportional to the intensity of light striking the cathode. (See Fig. 35.) The response of the tube for various wavelengths of light (colors) depends on the material used for the photocathode. The tubes can be made to approximate the response of the human eye to light, or they can be made sensitive to invisible ("black") light, either in the ultraviolet or infrared region. Phototubes are highly sensitive and accurate devices, used widely for precision laboratory light measurements and for critical control applications. They are also used in motion picture

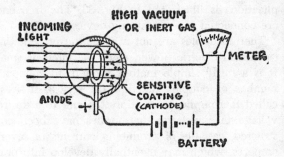

Fig. 35. Action of a Phototube

projectors to convert the light variations passing through the sound track of the film into the corresponding electrical variations of the original sound waves. Phototubes are *not* true generators of electricity, however, since they require an external voltage (emf) for their operation.

The Photovoltaic Cell. Photovoltaic cells (sometimes called, simply, **photocells**) generate their own emf when exposed to light and thus convert light *directly* into electric current. The action of these cells is similar to the thermoelectric effect we have discussed, in that it depends on the small **difference in potential** present between two conductors in contact (i.e., the **contact potential**). In the case of photovoltaic cells, however, the contact potential varies with the **illumination** of the contact, rather than its temperature.

The construction of one type of photovoltaic cell is illustrated in Fig. 36. The cell is a metallic "sandwich," consisting of three layers of different materials. One outside layer, called the **base**, is made of iron. The other outside layer is a thin gold or silver film, which serves as a semitransparent "window" for the incident light. The center layer, consisting of a thin, light-sensitive selenium alloy, is sandwiched between the two outside layers. The entire assembly is held together by a collector ring.

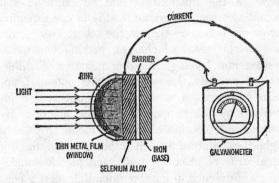
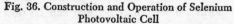

Fig. 36. Construction and Operation of Selenium Photovoltaic Cell

Because of its construction the device is sometimes called **barrier-layer** or **dry-disk** photocell.

When light shines on the translucent "window" of such a cell, an electromotive force which is directly proportional to the illumination is generated between the two outside layers. If a wire is connected between the layers, a current will flow through the circuit in proportion to the illumination (i.e., the emf generated) and the resistance of the external circuit. The current, and hence the light falling upon the window of the cell, may be measured by inserting a galvanometer into the external circuit. Because of their simplicity and relative accuracy photovoltaic cells are extensively used in photographic exposure meters.

The Photoconductive Cell. The electrical resistance of selenium and a few other substances depends on the intensity of illumination falling upon the material, an effect known as **photoconductivity**. This effect is made use of in **photoconductive cells**, which have similar but more limited applications than photovoltaic cells and phototubes. Like the phototubes, photoconductive cells require an *external emf*. An emf may be dispensed with, however, if the cell is made part of a resistance-measuring device (such as a Wheatstone bridge), which converts the resistance changes directly into a meter indication.

Basic Laws of Photoelectricity. Regardless of type, photosensitive surfaces (phototubes and photocells) obey the following empirical laws, when light falls upon them:

1. The number of electrons liberated from the surface, and hence the **current**, is **proportional to the intensity** of the incident light.

2. The maximum **kinetic energy** of each released electron (due to its velocity) is **independent of the light intensity**, but is directly proportional to the frequency (equals velocity/wavelength) of the light. (You will remember from basic physics that the wavelengths of light are measured in Angstroms and that the entire visible light spectrum of the rainbow colors extends from about 4000 Angstroms for violet to about 7400 Angstroms for red.)

3. From the second law it is apparent that there must be some frequency, known as the **treshold frequency**, below which the electrons will have insufficient kinetic energy to be liberated from the photosensitive surface. This is indeed so, and each type of surface material has its characteristic treshold frequency, below which no photoelectric current takes place.

ELECTRICITY FROM PRESSURE (PIEZOELECTRIC EMF)

When crystals of quartz, Rochelle salt and tourmaline are subjected to **mechanical pressure**, a displacement of charges takes place on the crystal faces, resulting in a **difference of potential** between them. Conversely, if a difference of potential (emf) is applied between the faces of such a crystal, the crystal will be slightly deformed in its dimensions. This is known as the **piezoelectric effect**, and it was discovered by the French physicists PIERRE and JACQUES CURIE in 1880. Although the effect is only temporary, while the pressure of potential difference lasts, it may be made continuous by alternating the pressure or emf. Thus, by alternating the pressure applied to the crystal faces between compression and tension, a continuous, alternating emf (varying between plus and minus) may be generated. Conversely, connecting the opposite sides of a piezoelectric crystal to a source of alternating emf (a.c.), continuous lengthwise **vibrations** may be set up in the crystal.

EXPERIMENT 11: Obtain a discarded crystal phono-cartridge and connect its output terminals to a sensitive galvanometer or millivoltmeter, as shown in Fig. 37. Place the cartridge flat on a table and strike it sharply, but not too heavily, with a hammer. Each hammer blow should produce a slight, momentary deflection of the galvanometer, indicating that an emf is generated between the crystal faces. (Note: Because of shock damping in a crystal cartridge the meter deflection may be too small to be noticeable. A better way is to place a fair-sized Rochelle salt crystal between metal plates, and connect the plates to a meter. The mere pressure of the hand on one of the metal plates will then be sufficient to produce an observable meter deflection.)

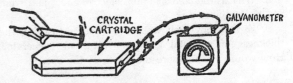

Fig. 37. A Rochelle Salt Crystal when Sharply Tapped Produces a Piezoelectric Emf That will Deflect a Galvanometer

Applications. The piezoelectric effect has many practical applications. Rochelle salt and other crystals are used widely in phonograph **pickups** and **crystal microphones** to convert sound (mechanical) vibrations into corresponding electrical variations. Since the output of a crystal pickup is only about one volt and that of a crystal microphone less than 1/100 volt, **electronic amplification** is necessary for either device. Piezoelectric crystals are also used in underwater **hydrophones** and in the piezoelectric **stethoscope**. In all these applications the crystal is used to generate an emf when subjected to mechanical vibration or sound.

As we have seen, connecting an alternating voltage to the opposite faces of a piezoelectric crystal sets up mechanical vibrations in the crystal. This **converse piezoelectric effect** also has many industrial applications. The entire field of **ultrasonics** is based on it. Here a high-frequency alternating voltage is applied to a **crystal transducer**, which then produces mechanical vibrations of the same frequency. The vibrations are called ultrasonic because their frequencies are beyond audibility. The ultrasonic sound waves may be used for reflection from submarines (sonar), to drill holes into defective teeth, or for **ultrasonic surgery**, to mention just **a** few applications.

Practice Exercise No. 4

1. Describe how work is involved in generating an emf and what happens to the work when current flows.

2. Explain why it is better to determine temperature by measuring the *emf* generated by a thermocouple rather than the thermoelectric *current*.

3. How can you increase the emf supplied by a thermocouple? How can you increase its capacity to deliver a thermoelectric current?

4. What are the basic laws underlying the photoelectric effect?

5. Distinguish between *photomissive, photovoltaic* and *photoconductive* cells and describe the characteristics of each.

6. Which types of photocells require electronic amplification and which do not?

7. Explain the role of certain crystals in phonograph pickups. Are all crystals piezoelectric? If not, which are?

8. What is the *converse* piezoelectric effect and what are its amplifications?

9. Explain how you would go about measuring the thickness of a metal using an ultrasonic generator. How would you measure the depth of the ocean?

SUMMARY

An **electromotive force** or difference of potential is generated by the **work** performed in separating electrical charges so as to create a **surplus** of electrons on the **negative** terminal (source) and a deficiency of electrons on the **positive** terminal (sink).

An emf is generated between the junctions of two dissimilar metals, when one is at a higher temperature than the other. This is called the **thermoelectric effect**. It is used in **thermocouples** and **thermopiles** in a variety of applications.

Photoelectric devices are classified as **photovoltaic** if the incidence of light on a **photosensitive** surface generates an emf; as **photoemissive** if electrons are **emitted** from the surface and must be collected by use of a **positive**, external voltage; and as **photoconductive** if the incidence of light changes the **internal resistance** of a substance. Photoelectric cells are used in a variety of control applications, for counting and sorting objects.

The **number** of electrons liberated from a photosensitive surface (and hence the current) is **proportional to the intensity** of the incident light. The kinetic energy with which each electron is emitted is **proportional to the frequency** of the incident light. Photoelectric emission cannot take place below the characteristic **treshold frequency** of a photosensitive material.

When crystals of quartz, Rochelle salt and tourmaline are subjected to mechanical pressure, a displacement of charges occurs on the crystal faces and a difference of potential (emf) results. Conversely, when an emf is applied between the faces of such a crystal, it becomes slightly deformed. This is known as the **piezoelectric effect**.

The piezoelectric effect is utilized in **crystal phono pickups, crystal microphones,** and similar devices to **convert sound waves** into corresponding **electrical variations**; it is used in **ultrasonic generators** to produce high-pitched, inaudible "sound" waves by the application of a high-frequency alternating voltage.

CHAPTER FIVE

ELECTRICITY FROM CHEMICAL ACTION
(ELECTROCHEMISTRY)

Chemical action is the second most important source of electricity, the most important being electromagnetism (generators). We are all familiar with the numerous useful jobs done by electric cells and batteries. They start our cars and power its electrical equipment; drive submarines; energize mobile radio equipment; power flashlights and fire flash bulbs; provide emergency lighting; and do many other things that could not be done as conveniently without batteries. To understand how chemical action can accomplish all this, we must make a few basic distinctions. Two dissimilar metal **electrodes** placed in a conducting chemical solution (called **electrolyte**) are capable of producing a potential difference (emf) between them. Such an arrangement is known as a **primary or voltaic cell** after its inventor, the Italian physicist ALESSANDRO VOLTA. Several cells connected together to provide either a greater emf (voltage) or a greater current capacity than a single cell, are known as a **battery.** (A single cell is often mistakenly called a battery.)

Instead of obtaining electricity from chemical action, we can reverse the procedure and *obtain chemical action from electricity.* Thus, placing a voltage on two metal electrodes in a special chemical solution will result in a current flow by **electrolytic conduction.** Moreover, the current flow results in **chemical decomposition** of the electrolyte and the electrodes, a process known as **electrolysis.** This action is made use of in depositing a layer of metal from one of the electrodes on top of the other (**electroplating**) and also in **reversible chemical cells** whose charge (emf) may be replenished by applying an external voltage to their electrodes. An important application of these reversible or **secondary cells** is the storage cell or **storage battery** (a group of secondary cells). All these interactions between chemistry and electricity are grouped together under the heading of **electrochemistry.**

PRODUCTION OF CHEMICAL EMF—
PRIMARY (VOLTAIC) CELLS

To clarify how an electromotive force can be generated by chemical action, let us perform a simple experiment:

EXPERIMENT 12: Obtain some strips of zinc and copper foil and a quantity of sulfuric acid (diluted 1:20). Wrap one end of a length of (#18) copper wire around a zinc strip and one around a copper strip. Connect the free ends of the wires to the terminals of a sensitive voltmeter. Now place or clamp the copper and zinc electrodes some distance apart in a clean tumbler which has been partially filled with the sulfuric acid. (See Fig. 38.) Note that the voltmeter will deflect as soon as both electrodes are immersed in the sulfuric acid electrolyte, and that an emf of about 1.1 volts will be indicated on the meter. Note further the formation of *bubbles* on the copper electrode of this simple primary cell *as long as current is flowing* through the external wires and the meter. If you disconnect one wire from the meter, interrupting the circuit, bubble formation will cease immediately.

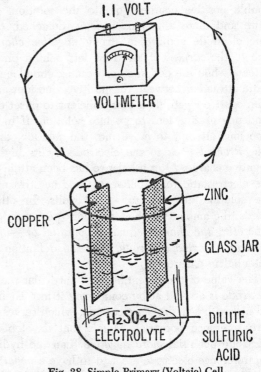

Fig. 38. Simple Primary (Voltaic) Cell

Basic Cell Action. Let us see what happens in the simple primary cell, illustrated in Fig. 38. The action here is essentially the same for all primary cells. Assume first that the wires are not connected and, hence, no current flows through the external circuit. All you have, then, are a copper and a zinc electrode immersed in sulfuric acid. The sulfuric acid (chemical symbol H_2SO_4), like all other acids, salts and bases, breaks apart or *dissociates* in a water solution into its constituent atoms. In this process, the sulfate (SO_4) molecules steal electrons from the hydrogen (H_2) atoms, so that the sulfate molecules become *negatively charged sulfate ions* (SO_4^-) and the hydrogen atoms become *positively charged hydrogen ions* (H^+). The **ionization** of an acid (or salt) in water thus leaves charged particles (ions) in the solution, which are capable of carrying an electric current. Such a conducting solution is called an **electrolyte**. (Pure water, oil, or a sugar solution are *not* electrolytes.)

When the highly active zinc electrode (chemical symbol Zn) is inserted into the acid solution, zinc atoms are thrust into the solution and the zinc slowly dissolves. Because of their high chemical activity, however, each of the zinc atoms readily parts with two electrons in its outer shell, leaving them behind on the metal electrode, while the *positively charged zinc ions* (Z^{++}, indicating a double positive charge) go into the solution. As more and more zinc ions become detached, the zinc electrode acquires a strong *negative charge* due to the *surplus* of electrons left behind on it. After a while the action stops because the **electrostatic attraction** between the positive zinc ions and the negative electrode is just sufficient to offset the tendency of zinc ions to go into solution. If more zinc ions attempt to move into the solution, they are *attracted back* to the electrode by its highly negative charge. After a balance has been attained, the electrostatic force between the zinc strip and the solution amounts to -0.76 volts. In other words, the amount of **chemical work** performed in *separating the charges* equals a *potential of -0.76 volts on the zinc electrode*. (See the definition of potential in Chapter 2.)

The value of the potential for a particular metal electrode is arrived at by comparing it in a chemical cell with a **hydrogen electrode**, which serves as **zero reference potential**. Metals that develop an electrode potential *more negative* than the hydrogen reference electrode are said to have a *negative* (—) potential and the electrode itself is labeled *minus* (—). Metals that develop an electrode potential *less negative* than that of a hydrogen electrode,

in contrast, are said to have a *positive* (+) potential and such an electrode is labeled *plus* (+). In this way, all the metals can be grouped by comparing their relative chemical activity with that of hydrogen, and this activity may be expressed in *volts* of electrode potential (emf). Such a grouping is known as the **electromotive series** of the metals; it is listed in Table IV for some typical metals. The value of the electrode potential in each case is a measure of the tendency of the metal's atoms to *lose electrons* and thus form **positive ions**, which go into solution. Keep in mind that an electrode always becomes **negatively charged** when this happens, regardless of the *sign* (+ or —) of the electrode potential. The sign of the potential is a relative matter with respect to the potential of a hydrogen electrode, which is arbitrarily set at zero volts, as we have explained. The electromotive series is a **continuous listing** of the chemical activity of the metals, with the *most active* element being on *top* and the *least active* one on the *bottom*.

TABLE IV
ELECTROMOTIVE SERIES OF THE METALS

Metal	Chemical Symbol	Electrode Potential (volts)
Lithium	Li	−3.02
Potassium	K	−2.92
Barium	Ba	−2.90
Sodium	Na	−2.71
Aluminum	Al	−1.67
Zinc	Zn	−0.76
Chromium	Cr	−0.71
Iron	Fe	−0.44
Nickel	Ni	−0.25
Tin	Sn	−0.14
Lead	Pb	−0.13
Hydrogen	H	0.00
Bismuth	Bi	+0.20
Copper	Cu	+0.34
Silver	Ag	+0.80
Mercury	Hg	+0.85
Gold	Au	+1.68

To return now to the rudimentary cell. As you can see from Table IV above, copper is much less active than zinc and develops an electrode potential of only 0.34 volts when it is inserted into the acid solution. The action is the same we have discussed for the zinc electrode, except that it is much milder. After a few positively charged copper ions

Cu^{++}) have been delivered to the solution, the copper electrode becomes sufficiently negatively charged by its surplus of electrons to stop the action. A balance is reached when 0.34 volts potential have been developed on the copper electrode, compared to a hydrogen electrode. Since the copper electrode is 0.34 volts *less negative* in the emf series than a hydrogen electrode, the potential is designated as *positive*, or +0.34 volts, and copper is called the *positive* electrode. Moreover, the **potential difference** (emf) developed between the copper and the zinc electrode is the *difference* between their individual potentials; that is, $+0.34 - (-0.76) = 1.10$ *volts*. A copper-zinc chemical cell—also known as **gravity cell**—thus develops a potential difference (emf) of 1.1 volts between its terminals. This emf is known as the **open-circuit voltage** of the cell, since it exists when the external circuit (wires) is open and no current is delivered. A sensitive voltmeter, which draws only a tiny current for its operation, will indicate this open-circuit voltage when it is connected between the terminals of a chemical cell.

Current Flow Through Cell and External Circuit. What happens when the electrodes of a cell are connected together externally by a wire (or load), as illustrated in Fig. 39? The equilibrium previously

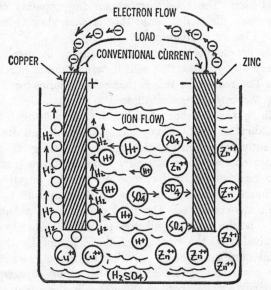

Fig. 39. Current Flow in Primary Cell

established at each electrode for the open-circuit condition is now upset, and the surplus of electrons on the zinc electrode starts to *flow through the external wire* to the copper electrode. The external electron flow thus takes place from zinc to copper,

or from — to +. Because of the Franklinian error which we described earlier, "conventional current" is said to take place from plus (+) to minus (—) through the external circuit, or from the copper to the zinc electrode. Unless otherwise stated we shall ignore this conventional current in this volume, and deal only with **electron flow**, whose direction is *opposite* to that of conventional current.

As more electrons flow over from the zinc to the copper electrode, more zinc dissolves and additional zinc ions (Zn^{++}) enter the solution. These zinc ions **displace** the hydrogen (H$^+$) ions in the solution near the zinc electrode and *drive them over to the copper electrode*. There is thus an *internal flow of hydrogen ions* within the cell from the zinc to the copper electrode. As each hydrogen ion arrives at the copper electrode, it *combines with an electron* reaching the electrode through the wire, thus forming **neutral hydrogen gas**. (That is, $H^+ + e^- = H^0$ or neutral hydrogen.) This neutral hydrogen gas begins to bubble out of the solution at the copper electrode, as you observed in the experiment (Experiment 12).

As the action continues, the copper electrode becomes almost completely *coated with hydrogen* gas bubbles, a condition termed **polarization**. The copper electrode now behaves more like one of hydrogen and the potential at this electrode drops to zero. The *difference* in potential between the copper and zinc electrodes is now only 0.76 volt (see Table IV), and the emf at the terminals therefore drops to this low value. Because of this lowered emf, the external current also drops, or equivalently, the **internal resistance** of the cell is said to have *increased*. (More about this later.) A substance, called **depolarizer**, is added in practical cells to prevent the formation of hydrogen gas by converting it into water. Even without depolarizer, the current continues to flow at the reduced emf until either the entire zinc electrode is "burned up" or various impurities caused by "local actions" have increased the internal resistance of the cell to a point where current stops flowing. The death of a cell is usually a messy affair, with both the copper and zinc covered with various impurities and gases, and the zinc almost completely eaten up.

Construction of a Dry Cell. Primary cells come in many shapes and types, consisting of various metal electrodes and acids or salts as electrolytes, depending on the application. The cell most widely used as a convenient source of portable electricity is the "dry cell," which actually is a hermetically sealed, non-spillable wet cell. (A completely dry cell would have no chemical action and hence no emf.) Dry cells are used for innumerable applications, such as small pencil flashlights, emergency lanterns, hearing aids, small radios, etc. The construction of a typical dry cell is shown in Fig. 40.

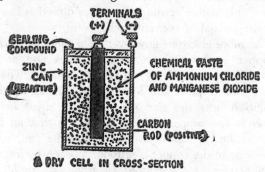

Fig. 40. A dry cell in cross section

As shown in the illustration, the entire zinc metal housing acts as the negative plate, while the carbon rod in the center serves as the positive plate. Screw terminals are provided on the zinc can and carbon rod to make external connections. The electrolyte is a chemical paste consisting of ammonium chloride mixed with manganese dioxide, which acts as "depolarizer" to take up the hydrogen. The cell is sealed at the top with a compound of pitch or wax. A dry cell provides an open-circuit voltage of about 1.5 volts, which drops considerably if any sizeable current is withdrawn. In operation, the metallic zinc delivers zinc ions (Zn^{++}) to the electrolyte and is consequently eaten away until the cell becomes useless. Moreover, after considerable use, the depolarizer no longer is able to take up the hydrogen as fast as it is released; as a result, the internal resistance of the cell increases and its open-circuit voltage drops until the cell can no longer deliver a useful current. No method has been found as yet to recharge an exhausted primary cell.

Batteries. A battery consists of a number of primary (or secondary) cells connected together. Cells may be connected either in series or in parallel. A parallel combination is made by connecting together all the negative terminals of the individual cells and also all the positive terminals, as illustrated in Fig. 41. In effect, this adds together the areas of the negative plates to make one large negative electrode and also those of the positive plates to make up one large positive electrode. Since the electrolytes are also added together, the resulting action is that of a single large cell with an internal resistance equal to that of a single cell divided by

the total number of parallel cells. Such a battery of parallel cells can supply a current that is the *product* of the individual cell current and the total number of cells. Note that the schematic circuit symbol for a parallel battery is the same as for a single cell.

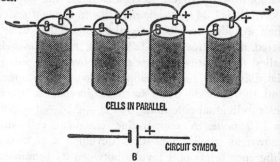

Fig. 41. Battery of Cells Connected in Parallel and Circuit Symbol

EXAMPLE: A No. 6 dry cell has an internal resistance of about 0.1 ohm and can deliver a maximum current of about 15 amperes. What is the internal resistance and the total current capacity of five No. 6 dry cells, connected in parallel?

SOLUTION: The internal resistance of five cells in parallel is *one-fifth* that of a single cell, or $\dfrac{0.1}{5} =$ 0.02 ohm. The total current-carrying capacity of the five cells is *five times* that of a single cell, or 5×15 amps $= 75$ *amperes*.

Series Connection. Batteries are more frequently made up by connecting cells in series than in parallel. The reason for this is that cells in series *multiply the emf* (potential difference) of an individual cell *by the number of cells*, thus permitting the buildup of fairly large voltages, which are frequently needed in practice. As shown in Fig. 42, a series connection is made by hooking a wire from the positive terminal of one cell to the negative terminal of the next in chain fashion, until all the cells are connected. The total emf of such a battery is the *sum* of all the individual emf's, or equivalently, the *product* of the cell emf by the total number of cells, provided all the cells have the *same* emf. The *total current-carrying capacity* of such a series battery, however, is the *same as that of a single cell*, because the total internal resistance has gone up by the same factor as the total emf. (We shall have more to say about that in the Chapter on Ohm's Law.) Note that the schematic circuit symbol of a series battery shows the addition of the individual emf's.

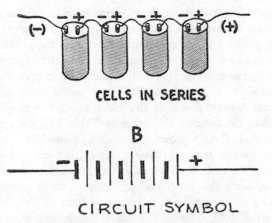

CELLS IN SERIES

B

CIRCUIT SYMBOL

Fig. 42. Series Connection of Cells and Schematic
Circuit Symbol

EXAMPLE: The five No. 6 dry cells mentioned in the last example are connected *in series* this time. If the emf of a single cell is 1.5 volts, what is the total emf of all five in series? What is the total internal resistance and the current-carrying capacity of the combination?

Solution: The total emf of identical cells connected in series is the product of the single-cell emf and the number of cells, or $5 \times 1.5 = 7.5$ volts. The internal resistance of the battery is the sum of all internal resistances; in this case $5 \times 0.1 = 0.5$ ohm. The total current-carrying capacity is the same as that of a single cell, or 15 amps. (You can show this by using Ohm's Law:

$$\text{total current} = \frac{\text{total voltage}}{\text{total resistance}} = \frac{7.5}{0.5} = 15 \text{ amps.})$$

ELECTROLYTIC CONDUCTION

Let us now reverse the earlier procedure and obtain some chemical action from electricity. Specifically, let us see what happens when an emf from a battery or other source is applied through two electrodes to a *liquid*. If this liquid is *pure water* almost nothing will happen, since water is a very poor conductor of electricity. To render it conductive we must add some charged particles that act as **carriers** of electricity. This is easily done by dissolving any acid, salt, or base in water. We have seen before that acids, salts, or bases break up into charged particles called **ions**, when in solution. You will recall that a **positive ion** is an atom that has lost one or more electrons, while a **negative ion** is an atom that has gained a surplus of electrons. For example, when hydrochloric acid (chemical symbol HCl) is placed into a water solution, it breaks up

or *dissociates* into positively charged hydrogen (H^+) ions and negatively charged chlorine (Cl^-) ions. This may be written in the language of chemistry

$$HCl \rightleftharpoons H^+ + Cl^-$$

where the double arrow signifies that the reaction may go both ways; that is, HCl molecules may dissociate into ions and these ions may also recombine into hydrochloric acid molecules. What happened in this reaction is that in the process of breaking up a chlorine atom steals an electron from a hydrogen atom, thus giving the chlorine atom a *negative* charge and the hydrogen atom a *positive* charge. The net effect is that the liquid now contains electric charges and can conduct electricity. Such a conducting liquid is called an **electrolyte**.

Electrolysis. Let us now place two chemically inert electrodes (*platinum* or *carbon*) into an electrolyte, say sulfuric acid in water, and connect a battery between the electrodes. (See Fig. 43.) The electrodes may both consist of the same metal, but they must be chemically *inert*, since we are interested only in the chemical action going on in the electrolyte and do not want to decompose the electrodes. To distinguish between the electrodes, the plate connected to the **positive** terminal of the battery is called the **anode**, while the plate connected to the **negative** terminal is called the **cathode**. Electrons *enter* the solution through the *cathode* and *leave it* through the *anode*, while conventional current flow is in the opposite direction.

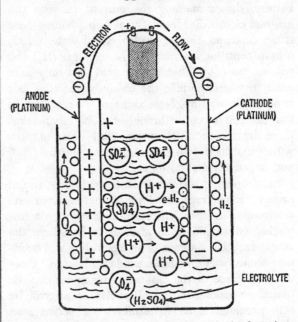

Fig. 43. Electrolysis of Water (Sulfuric Acid Solution)

The electrolyte of sulfuric acid (H_2SO_4) has dissociated into positive hydrogen (H^+) ions and negative sulfate (SO_4^-) ions, each of the two hydrogen atoms in the H_2SO_4 molecule having lost an electron to the sulfate group. The sulfate ion, consequently, has gained *two* electrons and is *doubly* charged, as indicated by the double minus sign in the symbol (SO_4^{--}). As soon as the battery is connected to the plates, the H^+ ions are attracted to the negative plate (cathode) and the SO_4^{--} ions are attracted to the positive plate (anode). At the same time free electrons flow out of the negative battery terminal into the cathode and enter the solution there. When an H^+ ion reaches the cathode, it combines with an electron to form a *neutral hydrogen atom.* Two hydrogen atoms make up a molecule of hydrogen gas (H_2), which bubbles up to the surface and escapes. The reaction may be written

$$2H^+ + 2e^- \rightarrow H_2 \uparrow \text{ (Hydrogen gas)}$$

where e^- stands for an electron and $\uparrow$ represents a gas.

At the *anode* another reaction takes place. Here each SO_4^{--} ion steals two *hydrogen* atoms from a *water molecule* (formula H_2O) to recombine to a neutral H_2SO_4 (sulfuric acid) molecule. The remaining oxygen atoms are set free and combine into molecules of oxygen gas (O_2), which bubble up to the surface. To balance the charges, some electrons are also set free in the process; these enter the anode and return to the positive terminal of the battery, thus sustaining the current through the external circuit. Evidently, what is happening here is the decomposition or **electrolysis** of water (H_2O) into its constituent elements, hydrogen gas (H_2) and oxygen gas (O_2). The sulfuric acid acts only as a **catalyst**, which permits the chemical reaction, but is not itself used up. Note also that current flow in the external circuit (through the battery) takes place by means of *electrons*, while current flow within the electrolyte is carried on by positive (H^+) and negative (SO_4^{--}) ions.

Electroplating. Electrolysis is a highly useful process, since it carries ions of matter from one electrode to another. The entire industry of **electroplating** one metal upon another is based on the simple fact that the ions of an electrolyte will transport atoms from a metallic anode and deposit them on the surface of the cathode. Any metal may be plated by making it the cathode in an electrolytic cell, *provided it is chemically more active than the metal to be plated upon it.* Chemical activity,

you will recall, depends on the place a metal holds in the **electromotive series** (Table IV). The higher the position of the metal in the series (or the greater its negative electrode potential in respect to hydrogen), the more active is the metal. It is possible, therefore, to plate practically any metal upon any other whose place in the electromotive series is above the former. Referring to Table IV, you can see that silver, for example, may be plated on lead, tin, nickel, iron, chromium, zinc, aluminum, and all other metals above it in the series. Copper or gold may also be plated upon these same metals, since all of them are more active than either copper or gold. You can make a simple experiment with copper plating which will show you the mechanics of the process.

EXPERIMENT 13: Obtain some bright, shiny iron nails, a strip of copper foil or heavy copper wire, a few copper sulfate crystals, two to three large (No. 6) dry cells, connecting wire and a fair-sized glass tumbler. Fill the tumbler partially with water and make a *saturated* copper sulfate solution by dissolving as many copper sulfate crystals as possible. (Adding a small amount of sulfuric or other acid will aid the reaction.) Connect the dry cells together *in series* to obtain an emf of 3 to 4.5 volts. Wrap (or solder) some connecting wire around the iron nail and the copper strip or wire. Connect the free end of the wire from the *iron nail to the negative terminal* of the dry-cell battery so that it becomes the **cathode.** Connect the free end of the wire from the *copper* electrode to the *positive* terminal of the battery, so that the copper serves as anode. (See Fig. 44.)

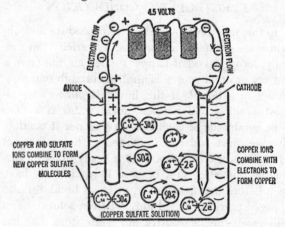

Fig. 44. Electroplating an Iron Nail with Copper

If you now place the copper and iron electrodes some distance apart in the copper sulfate solution, you will observe an immediate chemical reaction. Copper will start to deposit on the iron nail and after a few minutes the nail will be completely copper-plated. The longer the current lasts, the heavier will be the coating of copper on the nail. Do not try this too long, however, in order not to exhaust the dry cells. You will find that the plating will be much more *uniform* in thickness and adhere better, if you bend the copper into *cylindrical* form so that it encircles the iron nail, but does not touch it. If you should now *reverse* the battery connections, to make copper the cathode and iron the anode, you will find that the layer of copper on the iron nail may loosen and partially dissolve; you will not be able to deposit iron on the copper electrode, since only a less active metal can be deposited on a more active one. You can even substitute a clean *carbon* rod from a spent dry cell for the iron nail (as cathode) and you will see that the carbon becomes readily copper-plated.

Fig. 44 illustrates what takes place in our simple copperplating experiment. When you dissolved the copper sulfate crystals in water, a large number of positive copper ions (Cu^{++}) and negative sulfate ions (SO_4^{--}) were set free, in accordance with the reaction:

$$CuSO_4 \rightarrow Cu^{++} + SO_4^{--}$$

where the double plus ($^{++}$) and double minus ($^{--}$) signs indicate that *two electrons* are interchanged in the breakup of each copper sulfate molecule, so that the ions are doubly charged.

The positive Cu^{++} ions are strongly attracted to the negative iron electrode (i.e., the cathode) and move toward it. As each copper iron reaches the cathode it combines with two electrons, furnished by the battery, to form a neutral copper atom. Copper is thus deposited on the (iron) cathode, in accordance with this reaction:

$$Cu^{++} + 2e^- \rightarrow Cu \text{ (neutral copper)}$$

In the meantime, the negative SO_4^{--} (sulfate) ions are drawn over to the positive copper anode, where they combine with copper (Cu^{++}) ions dissolving from the anode. (The electrons freed by the Cu^{++} ions flow back to the positive terminal of the battery.) This combination of copper and sulfate ions results in the formation of *new molecules* of copper sulfate in accordance with the reaction

$$SO_4^{--} + Cu^{++} \rightarrow CuSO_4$$

The reaction shows that for each copper sulfate molecule that has given up a copper ion to the cathode, a new molecule is formed at the anode. The solution thus retains its full strength and the copper is simply *transferred* from the anode to the cathode.

Faraday's Laws of Electrolysis. The English chemist and physicist Michael Faraday discovered in 1832-33 two fundamental laws of electrolysis, which are still the basis of all quantitative calculations today. These laws may be formulated as follows:

1. The **weight** of any material deposited on the cathode during electrolysis is **directly proportional** to the **quantity of electric charge** passing through the circuit.

2. The passage of 96,500 coulombs of charge (called *one Faraday*) through an electrolytic cell deposits a **weight** (in grams) of *any* chemical element equal to the **atomic weight of the element divided by its valence.**

The first law appears fairly simple. It tells us that the weight of a substance deposited on the cathode (or, equivalently, liberated at the anode) is proportional to the quantity of electricity. The quantity of charge is usually measured in *coulombs*, which is the amount of electricity transported by a current of *one ampere flowing for one second* (ampere-second). (Equivalently, one ampere is a rate of flow of charge of one coulomb per second.) To obtain the total charge (in coulombs) that has passed through a circuit, therefore, you simply multiply the current (in amperes) by the time (in seconds). Sometimes a larger unit than the coulomb, called the **ampere-hour**, is used. An ampere-hour is the amount of charge transferred in *one hour* when the current is *one ampere.* (Since an hour contains 3600 seconds, one ampere-hour = 3600 ampere-seconds or *3600 coulombs.*)

EXAMPLE: A weight of 20 grams of a certain substance is deposited during electrolysis by the passage of 72,000 coulombs of charge. What weight of the substance will be deposited if a current of 2 amperes is maintained for 10 hours? What is the weight for a current of 4 amperes passing for 5 hours?

Solution: The weight in each case is the *same*, namely 20 grams. A current of 2 amps for 10 hours amounts to a charge of $2 \times 10 = 20$ ampere-hours. Since 1 ampere-hour equals 3600 coulombs, 20 ampere-hours $= 20 \times 3600 = 72,000$ coulombs or the *same* as the original charge. Finally, a current of 4 amps for 5 hours equals $4 \times 5 = 20$ ampere-hours, or again a charge of 72,000 coulombs.

Faraday's second law tells us that the same quantity of electricity will produce weights of *different* substances that are proportional to the *ratio* of the atomic weight to the valence for each substance. (This ratio is called the **chemical equivalent**.) Moreover, it states that a charge of *1 faraday* (96,500 coulombs) will liberate or deposit the *chemical equivalent* (atomic weight/valence) of any substance. You can see that the atomic weight must enter into it, since any substance is deposited atom by atom on the cathode, and the *number of atoms* in a gram depends on the atomic *weight*. Furthermore, each ion of the substance combines with one or more electrons to form a neutral atom of the substance. Thus, the copper ion (Cu++) with a valence of +2 requires *two* electrons to form a neutral copper atom; the hydrogen ion (H+) with a valence of +1, in contrast, requires only one electron to form a hydrogen atom. The greater the valence, therefore, the more electric charges (electrons) are required to form neutral atoms of the substance deposited. Hence, for a given total charge (total number of electrons) the weight deposited must be *inversely proportional* to the valence of the substance.

An example will further clarify the meaning of Faraday's second law of electrolysis.

EXAMPLE: Three electrolytic cells are connected in series with a battery, so that the same charge (current) passes through each of them. (See Fig. 45.) The first cell contains a solution of copper sulfate and deposits copper (atomic weight 63.5, valence +2) on the cathode. The second cell is filled with silver nitrate and deposits silver (atomic weight 107.9, valence +1). The third cell is filled with aluminum nitrate and deposits aluminum (atomic weight 27, valence +3). If one faraday of charge has passed through the circuit, what is the weight of the metal deposited at the cathode of each cell? What is the *weight per coulomb*?

Solution: According to Faraday's second law, 96,500 coulombs (1 Faraday) of charge will deposit the atomic weight/valence of any substance. Hence, we obtain for

Cell 1: $\frac{63.5}{2} = 31.75$ grams of copper;

for cell 2: $\frac{107.9}{1} = 107.9$ grams of silver;

and for cell 3: $\frac{27}{3} = 9$ grams of aluminum.

To obtain the weight deposited *for each coulomb* of charge, we must divide the figures above by 96,500. This turns out for copper 31.75/96500 or 0.0003294 gm/coulomb; for silver it is 107.9/96500

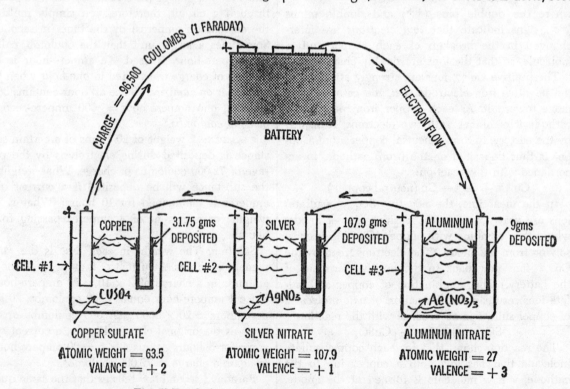

Fig. 45. Faraday's Second Law for Electrolytic Cells in Series

or 0.001118 gm/coulomb; and for aluminum it is 9/96500 or 0.0000933 gram/coulomb. The amount of material deposited for each coulomb of electricity, which we computed for the three elements above, is known as the **electrochemical equivalent** of the element.

The electrochemical equivalent of silver forms the basis for the legal definition of the **international ampere**. By an act of the U.S. Congress, the international ampere is specified as "the unvarying current, which, when passed through a solution of nitrate of silver in water in accordance with standard specifications, deposits silver at the rate 0.001118 gram per second." Since 1 ampere per second is 1 coulomb, this is the same as the electrochemical equivalent which we just computed.

Since the electrochemical equivalent of a substance is the weight deposited per unit charge (coulomb), you simply need to *multiply the electrochemical equivalent by the total charge* to obtain the weight of a substance deposited during electrolysis. If you do not know the electrochemical equivalent, you can compute it by the relation

$$\text{electrochemical equivalent} = \frac{\text{atomic weight}}{\text{valence} \times 96500}$$

Moreover, the total charge transferred is the product of the current (I) in amperes and the total time (t) in seconds, so that we can write the simple formula for Faraday's laws:

Total weight deposited = electrochemical equivalent times total charge or

$$\text{Total weight} = \frac{\text{atomic weight} \times \text{current} \times \text{time}}{\text{valence} \times 96,500}$$

$$= \frac{A\,I\,t}{V\,96500}$$

EXAMPLE: How much zinc (at. weight 65.38, valence +2) is deposited at the cathode of an electrolytic cell, if a current of 12 amps is passed through a solution of zinc salt for 20 minutes?

Solution: $W = \dfrac{A\,I\,t}{V\,96500} = \dfrac{65.38 \times 12 \times (20 \times 60)}{2 \times 96,500}$

$= 4.88$ *gms zinc.*

You can look up the atomic weights and valences of the elements in any handbook of chemistry or physics.

SECONDARY CELLS (STORAGE BATTERIES)

A secondary cell delivers current to a load by chemical action like a primary cell, but the chemical reaction in a secondary cell is *reversible*, permitting it to be restored to its original condition. All you have to do to restore or **recharge** a secondary cell is to pass a current through it in a direction opposite to that of its normal use or **discharge**. Combinations of secondary cells, called **storage batteries**, can furnish a relatively large amount of current for a short time, and since they can be recharged, they are a highly convenient source of power for mobile applications.

Lead-Acid Storage Cell. The most familiar type of storage battery is made up of **lead-acid storage cells**, each producing an emf of about 2 volts. A six-volt auto battery, thus, has three such lead-acid cells, while a 12-volt battery has six lead-acid storage cells. The case of a lead-acid cell (Fig. 46) is made of hard rubber or glass to prevent corrosion and acid leaks. The top of the case is removable and serves as support for the active plates (electrodes). To attain the maximum chemical action, a number of positive and negative plates are placed in the same electrolyte. The positive and negative plates alternate and are separated by *porous insulators* (wood or porous glass) which permit the electrolyte to pass through. Lead bars connect the plates of each polarity and serve as terminals on top of the case. A vent cap on the cover permits gases to escape. This cap may be removed to permit battery testing, refilling the electrolyte, or adding distilled water.

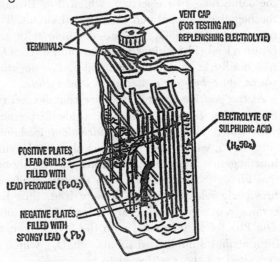

VENT CAP (FOR TESTING AND REPLENISHING ELECTROLYTE)

TERMINALS

ELECTROLYTE OF SULPHURIC ACID (H_2SO_4)

POSITIVE PLATES LEAD GRILLS FILLED WITH LEAD PEROXIDE (PbO_2)

NEGATIVE PLATES FILLED WITH SPONGY LEAD (Pb)

Fig. 46. Construction of Lead-Acid Storage Cell

The chemical action in a lead-acid storage cell involves reactions between lead compounds on the electrodes and the electrolyte of sulfuric acid (H_2SO_4). The *positive* plates consist of a grill or

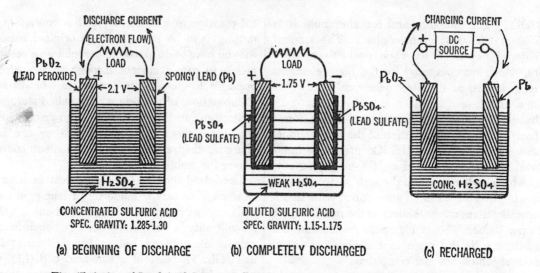

Fig. 47. Action of Lead-Acid Storage Cell During Discharge (*a* and *b*) and During Recharge (*c*)

lattice of a lead alloy coated with an active material of porous *lead peroxide* (chemical symbol PbO_2). The *negative* plates are a similar structure coated with *spongy lead* (Pb). Let us consider first the reactions taking place during *discharge* of the cell, when a current is being withdrawn from it. (See Fig. 47*a* and *b*.)

The sulfuric acid electrolyte is dissociated into positive H^+ ions and negative SO_4^{--} ions. At the *negative electrode* the spongy lead dissolves slightly, forming positive lead (Pb^{++}) ions and releasing at the same time two electrons, which flow through the negative terminal and the external circuit. The negative sulfate (SO_4^{--}) ions combine with the positive lead (Pb^{++}) ions into lead sulfate ($PbSO_4$), which adheres to the negative plates. The *negative plates*, thus, *become coated with lead sulfate.*

At the *positive* electrode a more complicated reaction takes place. The lead peroxide first reacts with water (H_2O) to form *quadrivalent* lead ions (Pb^{++++}), which have four plus charges, leaving four negative hydroxyl (OH^-) ions. The highly active Pb^{++++} ions then pick up the two electrons previously released by the spongy lead, thus becoming ordinary doubly charged lead (Pb^{++}) ions. The Pb^{++} ions now react with the SO_4^{--} ions to form again insoluble lead sulfate ($PbSO_4$), which is deposited on the positive plate.

The net result of these reactions is that *both plates* become coated with lead sulfate during discharge and the sulfuric acid is partially replaced by water, thus becoming *less dense*. The cell is completely discharged when both plates are covered

with lead sulfate and the electrolyte has become quite weak, as illustrated in Fig. 47*b*. Because of the consumption of sulfuric acid, its **specific gravity** drops from an initial value of about 1.30 (maximum) to about 1.15 (minimum), and the open-circuit voltage of the cell drops from about 2.1 volts at the start to about 1.75 volts for complete discharge. The best way to check whether the cell is charged is to measure the specific gravity of the electrolyte with a **hydrometer**. By sucking in some of the electrolyte into the hydrometer, the position of a "float" indicator will show the specific gravity, and hence the condition of the cell.

Recharge. The cell may be recharged by connecting the positive and negative plates, respectively, to the positive and negative terminals of a d-c source. Current now flows from the negative terminal of the source through the cell to the positive terminal, in a direction *opposite* to that of the discharge current. As a consequence, all the reactions previously described are reversed, and the lead sulfate on the positive plate is restored to lead peroxide (PbO_2) and the negative plate is restored to spongy lead. Moreover, the electrolyte returns to its original density and the open-circuit voltage again reaches about 2.1 volts. A cell may be restored to about 90 percent of its original condition during each recharging process and it may have a useful life of about two to three years.

The chemical reactions during charge and discharge that we have described may be conveniently summarized by a combined formula, which is in the form of a reversible chemical equation:

discharge

$$PbO_2 + Pb + 2H_2SO_4 \rightleftharpoons 2PbSO_4 + 2H_2O$$

↑   ↑   ↑ charge

(Pos. (Neg. (Elec-

Plate) Plate) trolyte)

This shows that during discharge both the positive plate (lead peroxide) and the negative plate (spongy lead) react with the sulfuric acid and become coated with lead sulfate ($PbSO_4$). The electrolyte is diluted by the formation of water. During recharging all reactions are reversed, as indicated by the arrow.

Edison (Nickel-Iron Alkali) Cell. Most of the faults present in the lead-acid cell are overcome by the Edison or nickel-iron-alkali cell. Because of its high cost the Edison cell is not as widely used in storage batteries as the lead-acid cell, though it is far superior to it. For the same ampere-hour capacity the Edison cell weighs only about half as much as the lead-acid cell, it is mechanically more rugged and not damaged by overloads and short circuits. The life of an Edison cell is substantially greater than its lead-acid cousin, though its operating efficiency is somewhat lower.

The positive plate of the Edison cell consists of nickel hydroxide, $Ni(OH)_2$ contained in pencil-shaped, perforated steel tubes, which in turn are inserted into a steel grid. The negative plate is of similar construction and contains perforated pockets that hold iron oxide, FeO, as active material. The electrolyte is a 21 percent solution of potassium hydroxide (KOH), to which a small amount of lithium hydroxide (LiOH) is added. Once the forming process is completed, the positive plate is essentially nickel dioxide (NiO_2) and the negative plate is made up of iron (Fe). During discharge, the nickel dioxide is chemically *reduced* to nickel oxide (NiO), while the iron is *oxidized* to iron oxide (FeO). During charge the reverse of this process takes place, with the electrolyte remaining unaffected in either case. The terminal voltage of a charged Edison cell is about 1.2 volts after a few hours of operation.

The Silver Cell. Another secondary cell has recently come into use, which is suitable in small rechargeable batteries for portable devices. This is the silver oxide-zinc cell, consisting of a positive silver oxide plate and a negative sheet of zinc. The electrolyte is a solution of sodium or potassium hydroxide. The silver cell has a high ampere-hour capacity per unit weight and is able to withstand relatively large overloads or short circuits. Its terminal voltage remains constant at approximately 1.5 volts.

Generation of Emf by Magnetic Action. We have discussed all the major sources of electric current in the last two chapters, except the most important one. This is the generation of an electromotive force by the **relative motion of a conductor in a magnetic field.** We shall see later (in the chapter on Electromagnetism) that every **current-carrying conductor is surrounded by a magnetic field,** and conversely, an **emf is generated in any conductor that moves through the lines of force of a magnetic field.** This latter action accounts at the present time for most of the electric power commercially produced. We shall defer the discussion of electric generators until we understand more fully the principles of electromagnetism.

Practice Exercise No. 5

1. A tingling, sour taste results when a clean copper penny and a clean dime are touched to opposite sides of the tongue. Explain.

2. Explain what happens when copper and zinc electrodes are immersed in sulfuric acid solution. What reactions take place when the two electrodes are connected by a wire?

3. Mercury and aluminum electrodes are placed in acid solution. (a) What emf do they generate? (b) What is the emf if the mercury is replaced by copper?

4. How is the electromotive series of the metals obtained?

5. Why does hydrogen form on the positive electrode of a primary cell? What are the effects on the emf generated and what can be done about it?

6. How does a *depolarizer* affect the internal resistance of a cell?

7. Explain the construction and action of a dry cell.

8. How would you make up a *battery* of dry cells to generate an emf of 15 volts?

9. If one cell has an emf of 2 volts and a maximum current capacity of 12 amps, how could you obtain 60 amps from five cells?

10. What is necessary to make a liquid an electrolyte? Name some.

11. Explain the *electrolysis* of water, including the reactions taking place at the *anode* and the *cathode*.

12. How would you plate a nickel spoon with silver? how a tin can with gold? Can you nickel-plate a silver spoon?

13. State Faraday's laws of electrolysis and explain them.

14. A current of 5 amps for 4 hours deposited 15 grams of a substance during electrolysis. How much of the substance will be deposited if a current of 15 amps is maintained for 6 hours?

15. (a) What is the *electrochemical equivalent* of

nickel, which has an atomic weight of 58.7 and a valence of $+2$? (b) If 50,000 coulombs of charge pass through a nickel chloride solution, how much nickel will be deposited?

16. Distinguish between *primary* and *secondary* cells?

17. Describe the reactions taking place in a lead-acid storage cell at the positive and negative plates during discharge and charge.

SUMMARY

A **primary** or **voltaic cell** consists essentially of two dissimilar metal electrodes placed in an electrically conducting solution (**electrolyte**). An emf is produced in such a cell by the separation of charge brought about by chemical action between the electrodes and the electrolyte.

The **electromotive force** generated by a primary cell depends on the relative positions of the electrodes in the **electromotive series of the metals**; the greater their separation in the series, the higher the emf. The emf can be calculated by taking the **algebraic difference** between the individual electrode potentials in the electromotive series.

A **dry cell** consists of a zinc metal housing, serving as **negative** terminal, a **positive** carbon electrode, and an electrolyte of ammonium chloride mixed with a manganese dioxide "depolarizer." A dry cell, when fresh, generates an emf of about 1.5 volts.

Chemical cells may be connected as **batteries**, either in series or in parallel. The **series connection** **multiplies** the **emf** generated by the number of cells, but permits a current no greater than for a single cell; the **parallel connection** multiplies the **current capacity** of a single cell by the number of cells, but generates an emf (voltage) no greater than that of a single cell.

Acids, salts and bases in liquid solution **dissociate** into electrically charged (positive and negative) **ions** that render the liquid electrically conductive; such a liquid is called an **electrolyte**. The breaking up into ions is known as **ionization**.

The passing of an electric current through an electrolyte (acid, base, or salt) results in its chemical decomposition, a process termed **electrolysis**. It takes place in an **electrolytic cell**.

In electrolysis the (electron) current enters the solution through the **negative electrode** (**cathode**) and leaves it through the **positive electrode** (**anode**).

Electroplating consists of passing an electric current through an electrolytic cell in which the **cathode** is made the **metal to be plated** and the **anode supplies the metal to be deposited**. The current will transport the anode metal and deposit it on the cathode, provided the **cathode is electrochemically more active than the anode and the electrolyte contains ions of the anode metal**.

Faraday's Laws of Electrolysis: 1. The **weight of any material deposited** or **liberated** during electrolysis is **directly proportional to the quantity of electric charge** passing through the cell.

2. The passage of 96,500 coulombs of charge (**1 Faraday**) through an electrolytic cell deposits or liberates a **weight** in grams of *any* chemical element equal to its **atomic weight divided by the valence**.

Electrochemical Equivalent of Element $=$

$$\frac{\text{Atomic Weight}}{\text{valence} \times 96500}$$

A **lead-acid storage** (**secondary**) **cell** has **positive plates of lead peroxide** (PbO_2), **negative plates of spongy lead** (Pb), and an **electrolyte of sulfuric acid** (H_2SO_4). The formula for charge and discharge is:

$$PbO_2 + Pb + 2H_2SO_4 \overset{\text{discharge}}{\underset{\text{charge}}{\rightleftharpoons}} 2PbSO_4 + 2H_2O$$

CHAPTER SIX

ELECTRICAL UNITS AND OHM'S LAW

We have discussed some aspects of electricity and have explored its sources. Now we are almost ready to make quantitative calculations in practical electrical circuits, which are so familiar in our everyday lives. But first we must accurately define the quantitative units of electricity, namely, **charge**, **current**, **voltage** (emf or potential difference) and **resistance.** The relationship between current, voltage, and resistance was discovered in 1828 by the German scientist GEORG SIMON OHM (1787-1854) and bears his name. **Ohm's law** makes possible 99 percent of all direct-current electrical calculations.

PRACTICAL ELECTRICAL UNITS

Charge. Electrical current consists of **charges** in motion. The smallest possible charge is that carried by an **electron**. Its charge is incredibly small, equaling about one-half billionth of an **electrostatic unit** (esu) of charge (more precisely, 4.8×10^{-10} esu). Even the electrostatic unit of charge (esu) is much too small a quantity of electricity for practical purposes; a much larger unit, the **coulomb**, is used. We have met both the esu and the coulomb before and you may recall that a *coulomb is the equivalent of three billion* (3×10^9) *electrostatic units of charge.* It may also be shown that a coulomb corresponds to the charge carried by a fantastic number of roughly *six billion billion electrons* (more precisely, 6.28×10^{18} electrons). You will understand why we prefer to calculate with coulombs rather than with electrons.

Current. Electric current is the **rate of flow** of electric charge. The unit of current, the **ampere** (named after the French scientist ANDRE M. AMPERE), represents a rate of flow of *1 coulomb per second.* Thus, if 10 coulombs pass a given point of a circuit in 5 seconds, the rate of flow of charge is $10/5 = 2$ coulombs per second, or 2 amperes. In general, **current equals charge per unit time**, a fact that may be expressed by the formula

$$I = \frac{Q}{t}, \text{ or } Q = I \times t$$

where I is the current, Q is the charge, and t represents time.

EXAMPLE: a current of 8 amperes passes through a wire for a period of 3 hours. What is the total charge transferred?
Solution:
$Q = I \times t = 8 \times (3 \times 60 \times 60) = 86,400$ coulombs.

Smaller units of current than the ampere (abbreviated amp.) are frequently used in practice. Thus, the *milliampere* (abbreviated ma) represents *one thousandth* of an ampere (1 ma $= 10^{-3}$ amp) and the *microampere* (abbr., μa) represents one millionth of an ampere (1 μa $= 10^{-6}$ amp). To convert *amperes to milliamperes*, simply move the decimal point *three* places to the *right* and to convert *amperes to microamperes* move the decimal point *six* places to the right. Conversely, to change milliamps to amps move the decimal point three places to the *left;* and to change microamps to amps, move it six places to the *left.*

EXAMPLE 1: Change 0.000357 amp into milliamps and into microamps.
Solution:
0.000357 amp $= 0.357$ milliamps $= 357$ microamps.

EXAMPLE 2: Change 7,584 microamps into milliamps and into amperes.
Solution:
1 microamp is 1/1000 milliamp $= 0.001$ ma. Hence, to convert microamps to milliamps, move the decimal point three places to the left. Thus 7,584 microamps $= 7.584$ milliamps

$$= 0.007584 \text{ ampere.}$$

Current Standards. It is desirable to have independent standards of electrical quantities, which are based on physical phenomena and can be arrived at by going through a certain experimental procedure in the laboratory. Until 1948 the so-called international units of electricity were commonly accepted, but in that year new **absolute units** were adopted, which differ only slightly from the international units. The **international ampere**, which is based on the **chemical effect** of an electric current, is defined as the *current that will deposit 0.0011183 gram of silver from a standard silver solution in one second.* The new **absolute ampere** is defined in terms of the **electromagnetic effect** of an electric current, which we shall describe in a later chapter. The new definition makes the absolute ampere somewhat larger than the international ampere, so that

1 absolute ampere = 1.000165 international ampere

and

1 international ampere = 0.999835 absolute ampere

It is very doubtful that you will ever have to worry about the difference between the absolute and the international ampere.

Voltage. As we shall see later on, the term "voltage" is a catchall for a variety of electrical concepts. Voltage may stand for the **electromotive force** or **potential difference** between the terminals of an electric source. (The symbol E is often used for this application.) As you know, the **open-circuit** voltage of such a source drops to a lower value, called **terminal voltage**, when a current is withdrawn from the source. (The symbol V is generally used to designate terminal voltage.) Finally, when a current flows through a resistance it develops a potential difference between its ends, which is referred to as a **voltage drop** (symbol V) to distinguish it from the voltage *rise* taking place in a battery or other source. All these varying concepts, with which we shall become more familiar, are designated as **voltage** and are measured in units of **volts**.

For the purposes of definition, we shall recall the **work concept** of voltage, which makes the potential difference synonymous with the **work done in transporting a unit charge** from one electrical level (potential) to another. In accordance with this concept we define the **potential difference** between two points in a circuit as *one volt* if *one joule of work* must be *expended* to move a *positive charge of one coulomb* from the point of *low* potential to the point of *high* potential. Instead of moving the charge *against* the force of the field from a low to a high potential, we can let it be *repelled by the field* from a point of high potential to a point of lower potential, in which case work will be done *by* the charge. Again, the potential difference is *one volt*, if the charge *performs one joule* (10^7 ergs) *of work* in moving from the point of *high* potential *to* the point of *low* potential.

The work or **energy concept** of voltage is useful in another way, as we shall see more clearly later on. When a current (i.e., charges in motion) flows through a circuit, the charges perform a certain amount of *work* in moving from a point of *high* potential (at one terminal of the electric source) to a point of *low* potential (at the other terminal of the source). The energy for doing this work must be supplied, of course, by the source of electricity. Moreover, since the energy expended must *equal* the energy supplied, it follows that the sum of all the potential drops (voltage drops) around the entire circuit must equal the emf of the source. This is an important fact to remember.

As in the case of current, the prefixes *milli-* and *micro-* are frequently used to designate smaller units of voltage. Thus, *one millivolt* (abbreviated *mv*) equals *one thousandth* of a volt (1 mv = 10^{-3} volt) and *one microvolt* (abbreviated *μv*) equals *one millionth* of a volt. As before, to change *volts to millivolts* move the decimal point *three* places to the *right;* and to change *volts to microvolts* move the decimal point *six* places to the *right.* Conversely, move the decimal point three or six places to the *left,* if you want to change millivolts or microvolts, respectively, to volts. In addition to these units, there is also a larger unit of voltage, called the **kilovolt** (abbreviated kv), which represents 1000 volts. Consequently, to change *volts to kilovolts* move the decimal point *three places to the left;* and to change *kilovolts to volts,* move it *three places to the right.*

EXAMPLE: Change 0.00045 kilovolt into volts, millivolts, and microvolts.

Solution: 0.00045 kv = 0.45 volts = 450 millivolts = 450,000 μv.

Voltage Standards. The definition of volts as *joules per coulomb* automatically gives us the presently adopted **absolute** volt. The absolute volt is 0.999670 of the old *international volt,* which was defined as the emf required to drive a current of one international ampere through a resistance of one international ohm. As laboratory standards of voltage, stable chemical cells are used, which maintain their emf over long periods of time. One of these **standard cells** is the **Weston normal** or **saturated cell,** which maintains an emf of 1.01865 volts at 20 degrees centigrade (68° F), provided no more than 50 microamperes current are drawn from it. Another laboratory standard cell is the **unsaturated cadmium cell,** which has an emf of 1.0192 volts.

Resistance. We have said that the opposition which free electrons encounter in moving through a material (conductor or insulator) is called the **resistance** of the material. (Conversely, the *ease* with which electrons move through a material is known as the **conductance** (symbol G) of the material.) Resistance (symbol R) is akin to mechanical friction, since it is caused by collisions between free electrons and the atoms of a material. The atomic or crystal structure of a material, therefore, determines its *inherent* resistance per unit length and area, which is sometimes called specific resistance or resistivity. You can calculate the resistance of a conductor if you know its resistivity, its length and its cross-sectional area, as we shall see presently.

Resistance is measured in units of ohms. The ohm is defined as the resistance of a conductor across which there is a potential drop of 1 volt, when a current of 1 ampere flows through it. If the current and voltage are in *absolute* units, then this statement defines the absolute ohm. The absolute ohm is 0.9995 as large as the *old international ohm* (1 international ohm = 1.000495 absolute ohms), which was defined as "the resistance offered by a column of mercury of 14.521 grams mass and 106.3 cm length, kept at the temperature of melting ice." If this definition sounds somewhat abstruse, you can get a better idea of an ohm by considering that a 1000-ft long copper wire, 0.1 inch in diameter (No. 10 American Wire Gage), has a resistance of 1 ohm; so does a copper wire 2.4 feet long and 0.005 inch in diameter (No. 36 gage). (We shall discuss *resistance in terms of the heat it liberates* in the chapter on Electric Power and Heat.)

The ohm is frequently abbreviated in numerical examples and on diagrams by the Greek letter *omega* (Ω or ω). Large values of resistance are expressed in megohms (1 megohm = 10^6 or 1 million ohms) and in kilohms (always abbreviated K = 1000). To change kilohms to ohms, move the decimal point three places to the right; and to change megohms to ohms move it six places to the right. (To do the converse, move the decimal point left the same amount.) Thus, a resistance of 500K equals 500,000 ohms or 0.5 megohm. The term microhm also is occasionally used to designate a millionth of an ohm (1 microhm = 10^{-6} ohm). A resistance of 0.005 ohms, for example, equals 0.005 × 10^6 or *5000 microhms*.

It is sometimes convenient to speak of the conductance of a wire rather than its resistance. Conductance is the *reciprocal* of resistance ($G = \dfrac{1}{R}$); and to indicate this inverse relationship, the units of conductance are *mhos* (ohms spelled backwards). A *millionth of a mho* is called the micromho (1 μmho = 10^{-6} mho).

RESISTANCE OF WIRES AND RESISTORS

Every material offers some resistance to the flow of electric current. Conductors have a relatively low resistance; insulators have a very high resistance. Moreover, the resistance of a wire conductor is affected by a number of factors, including the inherent **resistivity** of the wire, its length and cross section, and also by the surrounding **temperature**. To calculate resistances we must become acquainted with the interrelation between these factors. Also, it is frequently necessary to insert a fairly large resistance into a circuit without taking up too much space. A thousand feet of No. 10 copper wire is obviously not a practical way to obtain a resistance of one ohm. Specially designed resistors, consisting of high-resistance wire, carbon, or a composition material, are available to serve as **lumped resistances** of small dimensions.

Resistance of Wire Conductors. It has been found experimentally that the resistance of a wire **increases directly with its length** and **decreases in direct proportion to the area of its cross section** (i.e., its thickness). The resistance of a wire also depends on its inherent **resistivity** (symbol ρ, pronounced rho), where **resistivity is defined as the resistance of a wire sample of unit length and unit cross section.** These experimental findings may be quantitatively expressed by the formula

$$R = \rho \frac{L}{A}$$

where R is the resistance in ohms, ρ is the resistivity, L is the length of the wire and A is its cross-sectional area.

Two systems of units are in use to express the length, area, and resistivity of a wire. One system of units, mostly in use in Europe, expresses the *length in centimeters* (cm) and the area of the *cross-section in square centimeters* (cm^2), in which case the resistivity (ρ) comes out in *ohm-centimeters*. The other system, used by electricians in the United States, is based on the *length* of the wire expressed in *feet* and the cross-sectional *area* expressed in *circular mils*. The resistivity in this case is called the *ohm-circular mil per foot*, which is abbreviated in the electrician's language to *ohms per mil-foot*. Fig. 48 illustrates the two systems. Part (*a*) shows a wire 1 cm in length and 1 cm^2 in cross section whose resistance in ohms is equal to its resistivity, expressed in ohm-centimeters. For such a specimen of *copper* wire, the resistivity turns out to be

1.724×10^{-6} ohm-cm, and hence the resistance is 1.724×10^{-6} ohm. Part (*b*) shows a wire 1 ft in length and 1 circular mil in cross section. (Diameter is 1 mil = 0.001 inch.) For copper the resistivity is about 10.4 ohms per circular mil-foot at ordinary room temperature, indicating that the resistance of this specimen is approximately 10.4 ohms.

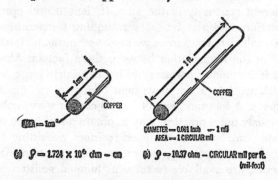

(*a*) $\wp = 1.724 \times 10^{6}$ ohm — cm   (*b*) $\wp = 10.37$ ohm — CIRCULAR mil per ft.
(mil-foot)

Fig. 48. Units of Resistivity (a) ohm-centimeter, (b) ohm-circular mil per foot (mil-foot)

A **circular mil** is a convenient unit for expressing **circular areas**. As you may know, the ordinary mil is a thousandth of an inch (1 mil = 0.001 inch). To find the cross-sectional area of a wire in circular mils, simply express its *diameter in mils* and *square* this number (i.e., circ. mils = (mils)² .) This unit avoids the use of the "π-factor." Thus, a wire of 0.08 in. diameter, has a cross section of $(0.08 \times 1000)^2 = (80 \text{ mils})^2 = 6,400 \text{ circular mils}$.

EXAMPLE 1: What is the resistance of 100 meters of No. 16 aluminum wire (diameter 0.13 cm.) having a resistivity of 2.63×10^{-6} ohm-cms?

Solution: The cross-sectional area of the wire

$$A = \frac{\pi d^2}{4} = \frac{3.14 \times (0.13)^2}{4} = 0.0133 \text{ cm}^2$$

Hence, the resistance

$$R = \rho \frac{L}{A} = \frac{2.63 \times 10^{-6} \times 100 \times 10^2}{0.0133} = 1.98 \text{ ohms}$$

EXAMPLE 2: Compute the resistance of 1000 feet of No. 10 copper wire (diameter 0.102 in.) with a resistivity of 10.4 ohms/mil-foot.

Solution: Express the area in circular mils first by squaring the diameter in mils. Thus, 0.102 in. = 102 mils; hence the area is $(102)^2 = 10,400$ circular mils. The resistance, therefore, is

$$R = \rho \frac{L}{A} = \frac{10.4 \times 1000}{10,400} = 1 \text{ ohm.}$$

(Thus, the resistance of 1000 ft of No. 10 copper wire is 1 ohm, which is a good value to remember.

(The exact value, as given in standard wire tables, is 0.9989 ohm.)

Kinds of Wires. Wires come in various types and sizes depending on use (indoors, outdoors, fixed, mobile, etc.) and current-carrying capacity. For electrical purposes most wires have at least two things in common: they are **round** and they are made of **copper**. Copper is practically always used because of its excellent **conductivity** (low resistivity). The insulation around solid copper wire depends on the application. For house wiring and indoor uses, the wire is usually covered with **rubber** and a layer of fabric on the outside, and it is run in a **cable** or **conduit** containing many insulated wires. For outdoor lines rubber is not used, the insulation generally consisting of several layers of **weatherproofed fabric braids**.

The type of wire which you are most likely to encounter for the usual household applications is **flexible** cord, consisting of a number of **stranded copper** wires twisted together into a single conductor. Fig. 49 illustrates three types of flexible cords with different insulations. The lamp cord consists of two insulated stranded-wire conductors enclosed by a cotton or rayon outer braid. Even more popular is rubber-covered flexible cord, consisting of a parallel pair of stranded copper conductors. These two types of cords are used for 90 percent of all small household appliances, lamps, portable radios, TV sets, etc. Devices which develop a considerable amount of heat, such as electric irons, toasters, heaters, etc., are connected by means of ironing cord, which uses copper conductors covered with rubber, an intermediate layer of fireproof asbestos and an outer covering of fabric.

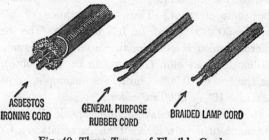

ASBESTOS IRONING CORD   GENERAL PURPOSE RUBBER CORD   BRAIDED LAMP CORD

Fig. 49. Three Types of Flexible Cords for Household Uses

Wire Sizes and Tables. The thickness of a wire determines its resistance for a given length and, hence, also its current-carrying capacity in a circuit. The diameter (thickness) of copper wire is specified by standard gauge numbers, known in the

United States as *American Wire gauge* (AWG). The *thicker* the wire, the *smaller* is its gauge number; the *thinner* the wire, the *greater* is its gauge number. Sizes of solid copper wire start at No. 0000 for a diameter of 460 circular mils and run all the way to gauge No. 40 for a wire of 3.145 circ. mils diameter. Fig. 50 illustrates the relative sizes of gauge numbers up to #18.

| 18 | 16 | 14 | 12 | 10 | 8 | 6 | 4 | 2 | 0 | 00 |

Fig. 50. Relative Thickness of Wires (American Wire Gauge)

Wires used in buildings run from gauge No. 12 to about gauge No. 18. For house and building wiring you are not permitted to use anything smaller than No. 14 gauge, which is rated at 15 amperes current (maximum) for rubber-covered wire in a conduit or cable. With the large power demand for air conditioners, television, and many other appliances, most modern houses actually use either No. 12 or No. 10 wire, rated at 20 and 25 amperes, respectively. Thinner wires heat up unduly and their relatively high resistance leads to a considerable loss of line voltage due to the **voltage drop** in the wire, which in turn causes decreased efficiency of appliances connected to the wire. No. 16 and No. 18 gauge wire is used inside lighting fixtures and in flexible cords. Flexible cords of stranded copper wire have somewhat larger current-carrying capacity than solid wire in cables or conduits. Thus, No. 16 flexible cord may carry up to 15 amperes current, and No. 18 wire up to 10 amps. It is most important to choose the right size wire for efficient and safe wiring. In a later chapter you will learn how to figure the current requirements of various household appliances, which will enable you to choose the right size of wire.

In the appendix of this volume you will find AWG copper wire tables, which will give the gauge numbers, diameter, cross section, resistance per 1000 feet, and other useful data for standard copper wire. It is, however, quite convenient to memorize some data. In our last example we computed that 1000 feet of No. 10 wire (diameter 100 mils, approximately) have a resistance of about 1 ohm. The cross-sectional area of No. 10 wire we found to be roughly 10,000 circular mils. Keeping these facts in mind and observing the regularities in the wire tables, you can figure the size and resistance of any wire roughly by means of the following rules:

1. The *resistance* of copper wire *doubles* and its cross section is cut in half each time you go up three gauge numbers. Conversely, the resistance drops to half and the cross section doubles, each time you go *down* three gauge numbers. (For example, No. 13 wire has a resistance of about 2 ohms per 1000 ft and a cross section of roughly 5000 circular mils; No. 7 wire has a resistance of 0.5 ohm and a cross section of about 20,000 mils.)

2. The *diameter* of the wire is multiplied by 1.41 as the gauge No. drops by a factor of 3; and *doubles*, when the gauge number goes down by 6. If the gauge number *goes up* by 3, divide the diameter by 1.41, and if it goes up by 6, the diameter is one-half. (For intermediate gauge numbers, take the square root of the area).

3. To obtain the resistance of a wire *one higher* in gauge number than that of a known resistance, *multiply the known value by 1.26;* to obtain its cross section *divide the known cross section by 1.26.* Do the converse to find the resistance and cross section of a wire *one lower* in gauge number than one of known resistance and area. (For example, No. 11 wire has a resistance of $1 \times 1.26 = 1.26$ ohms per 1000 ft and a cross section of $10{,}000/1.26 = 8000$ circular mils; No. 9 wire, in contrast, has a resistance of $1/1.26 = 0.8$ ohm per 1000 ft (approx.) and a cross section of $10{,}000 \times 1.26 = 12{,}600$ circular mils, roughly.)

EXAMPLE: Compute the approximate resistance per 1000 ft and the cross-sectional area of No. 17 copper wire.

Solution: Resistance of No. 10 wire is 1 ohm and the cross section is 10,000 circular mils. The resistance of *No. 16 wire*, therefore, is about 4 ohms and its cross section is 2,500 circular mils. Hence, for *No. 17 wire*, the resistance is $1.26 \times 4 = 5.04$ *ohms* (roughly) and the cross section is $2{,}500/1.26 = 2{,}000$ circular mils, approximately.

Resistance of Conductors in Series and in Parallel. According to the formula $R = \rho \dfrac{L}{A}$ the resistance of a conductor is **directly proportional to its length** and **inversely proportional to its cross-sectional area**. Consequently, if we join a number of identical conductors of equal length and cross section *end to end*, the resistance of a single conductor will be *multiplied* by the number of conductors thus joined in series (See Fig. 51.) For the example of three conductors in series, illustrated in Fig. 51a, the resistance of the combination is *three times* the resistance of a single conductor. In contrast, if we join several identical conductors side-by-side, or in parallel, the cross-sectional area goes up in direct proportion to the number of conductors and, hence, the **resistance of the parallel combination** is that of a **single conductor divided by the number of conductors in parallel**. Again, for the example of three identical conductors in parallel, illustrated in Fig. 51*b*, the resistance of the parallel combinations is *one-third* that of a single conductor.

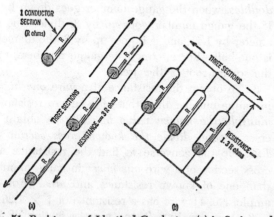

Fig. 51. Resistance of Identical Conductors (a) in Series and (b) in Parallel

If the conductors are *not* identical, you must *add the resistance* of each to get the total resistance of a *series* combination; and you must *add the conductance* (1/resistance) of each to obtain the total conductance (1/total resistance) of a **parallel** combination. As we shall prove in the next chapter, these relations hold *not* just for wire conductors, but *in general* for resistances in series or in parallel. More about this later on.

EXAMPLE: Sixteen wire conductors with a resistance of 2 ohms each are joined to make four series combinations of four conductors each. The four series combinations are then joined in parallel to make up a single composite conductor. What is its total resistance?

Solution: Each series combination has a resistance of $4 \times 2 = 8$ ohms. Joining these four 8-ohm "conductors" in parallel, we obtain a resistance of $8/4 = 2$ ohms, or exactly the same as that of a single conductor. The only advantage here is that the composite conductor can carry a far **heavier current** than each of the conductors.

Resistivities of Different Conductors. The resistivities of various metals and, hence, their ability to conduct electricity differ widely. Although copper is the chief metal for electrical conductors, other metals are employed frequently for special appli-

cations. For example, metals of high resistivity, such as Constantan, Manganin, and Nichrome, are used for heating applications and for resistors. Table V below lists the resistivities of some commonly used metals.

TABLE V

RESISTIVITIES OF METALS
(AT 20°C)

Material	Resistivity	
	ohm-cm $\times 10^{-6}$ (Microhm-cm)	ohm-circular mil/ft (ohms per mil-foot)
Aluminum	3.2	17
Copper	1.724	10.37
Iron	10 to 12	58.8
Silver	1.65	9.8
Nickel	8.7	51
Zinc	5.9	35.5
Constantan (Cu + Ni)	49	296
Manganin (Cu + Mn + Ni)	45	270
Nichrome (Ni + Cr + Fe)	112	675

Effect of Temperature on Resistance. Note in table V that we have specified the resistivity at a *temperature of 20°C* (68°F). This is necessary because the **resistance of pure metallic conductors increases with temperature.** A simple relation gives the law of increase of resistance with temperature:

$$R_t = R_o (1 + \alpha t)$$

where R_o = original resistance at the *reference temperature* (usually 20°C or 68°F).

R_t = final resistance at the higher temperature.

t = the *increase* in temperature (i.e., final temperature less original temperature).

and α = *temperature coefficient* of resistance.

The temperature-resistance coefficient of most metallic conductors averages about 0.004 per degree change of centigrade temperature. Certain alloys, such as Manganin and German silver, have extremely small temperature coefficients (about 0.00001 to 0.0004), which makes them useful for the construction of high-precision resistors with stable resistance values. Various **semiconductors** actually exhibit a *negative* temperature-resistance characteristic, which means that their **resistance decreases** with an increase in temperature. This characteristic is used to make temperature-sensitive resistors (called **thermistors**) to compensate for the rise in resistance of other components, and for use in control and measuring applications. Some thermistors drop in resistance as much as 10,000,000:1 when heated over a range of about 500°C.

EXAMPLE: A wire resistor has a resistance value of 50 ohms at 20°C (68°F) and a temperature coefficient of 0.004. What is the value of the resistor when the surrounding temperature is 100°C?

Solution: The *change* in temperature, t = 100 − 20 = 80 degrees

Hence the new resistance $R_t = R_o (1 + \alpha t) =$
50(1 + 0.004 × 80)
= 50 × 1.32 = 66 ohms.

Types of Resistors. Resistors are compact sources of "lumped" resistance. They come in a great variety of types depending on usage, resistance, power rating, required precision (tolerance), etc. Resistors range in size from very tiny (½-inch long) rod types for low power applications (½ to 2 watts) to huge "stick" structures used as high-power "ballasts" and for starting large motor-generators. Fig. 52 illustrates a few types of "fixed" resistors of constant resistance.

The carbon resistors (*a*), consisting of a rod of compressed graphite embedded in binding material, are very popular for low power applications (radio, electronics) requiring not too great precision. They come in resistance values below 1 ohm to several megohms, and have tolerances from 5 to 20% of the indicated value. The value and tolerance of the resistor are generally indicated by colored bands around its body, in accordance with a standard color code (see Appendix). Metal film resistors (*c*) are made by spraying a thin layer of a metal on a glass rod. For higher powers and greater precision, wire-wound resistors (*d*) are generally used. These are constructed by winding resistance wire of a low temperature coefficient (Nichrome, Manganin, German Silver) on a flat mica card, or on a porcelain or bakelite form. Precisions of about ± 1% tolerance of the indicated resistance value are possible.

Variable Resistors. Variable or adjustable resistors are generally either of the carbon type for low power applications or of the wire-wound type for greater power needs. (See Fig. 53.) The carbon types (*a*) are usually circular in shape and consist of a sliding contact attached to a rotating shaft, which rotates the movable contact over the carbon

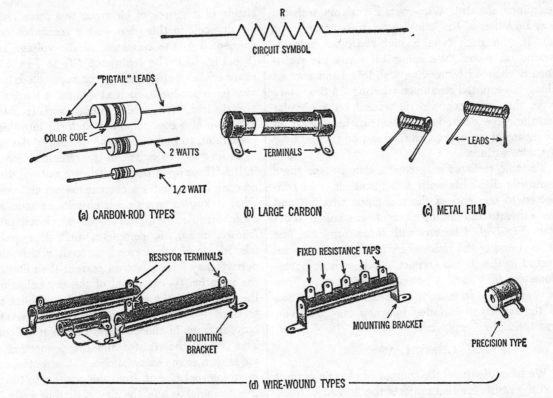

Fig. 52. Various Types of Fixed Resistors: (a) Carbon-rod types; (b) large carbon; (c) metal film; (d) wire-wound types

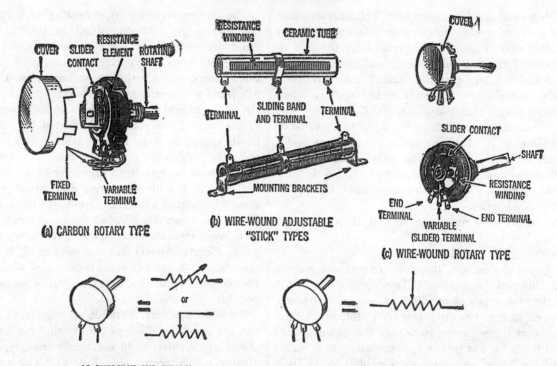

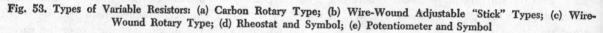

(a) CARBON ROTARY TYPE   (b) WIRE-WOUND ADJUSTABLE "STICK" TYPES   (c) WIRE-WOUND ROTARY TYPE

(d) RHEOSTAT AND SYMBOL   (e) POTENTIOMETER AND SYMBOL

Fig. 53. Types of Variable Resistors: (a) Carbon Rotary Type; (b) Wire-Wound Adjustable "Stick" Types; (c) Wire-Wound Rotary Type; (d) Rheostat and Symbol; (e) Potentiometer and Symbol

resistance element. Wire-wound variable resistors may be either of the "stick" type (*b*) or rotary type (*c*). In the stick type a high-resistance wire is wound on a straight ceramic tube and the resistance is changed by moving a sliding band contact along the exposed resistance winding. In the rotary type, the resistance wire is wound on a circular form and the desired resistance can be tapped off by means of a contact arm that can be rotated over the wire surface.

Variable resistors may have either two or three terminals. Resistors with *two* terminals, one connected to one end of the resistance winding and the other to the sliding contact, are called rheostats (Fig. 53*d*). Resistors with *three* terminals, one at *each* end of the resistance winding and one connected to the sliding contact, are known as potentiometers (Fig 53*e*). A potentiometer permits "tapping off" the voltage applied across it in proportion to the resistance included between one fixed end and the sliding contact.

OHM'S LAW

We have discussed the sources of electric current and the resistance that opposes the flow of current. Let us now consider a simple **electric circuit**, con-

sisting of a source of electromotive force (voltage) —a dry cell in this case—and a resistance or **load** connected to the terminals of the voltage source. (See Fig. 54.) The resistance (*R*) in Fig. 54 may represent an actual resistor or some electrical device (called a **load**), such as a lamp, a toaster or an electric iron, from which useful work is obtained. We have also connected a *switch* (S) into this simple circuit, to permit *opening* or *closing* the circuit.

As long as the switch in the circuit of Fig. 54 is in the UP or *open* position (shown dotted), there is no complete path for a current to flow and we have what is known as an **open circuit**. As soon as the switch is placed in the DOWN or closed position (shown solid), a complete, unbroken pathway (**closed circuit**) is formed through which electric current may flow. **Electron current** then flows from the negative (—) terminal of the dry cell, through the switch and the resistance load, and back to the positive (+) terminal of the dry cell. (Conventional current flows in the opposite direction, of course.) The switch, the resistor and the connecting wires are known as the **external circuit**. Current also flows in an **internal circuit**, from the positive to the negative terminal *inside* the dry cell, thus completing the electrical path. In such a circuit **electrical**

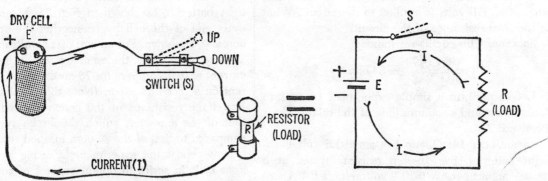

ACTUAL CIRCUIT   SCHEMATIC DIAGRAM

Fig. 54. Ohm's Law in Simple Electric Circuit

energy is supplied to the terminals of the dry cell by the chemical action inside the cell. This energy is then *expended in the external circuit,* either by heating up the connecting wires and the resistor, or by *performing useful work in a load.* The action continues as long as the voltage source can maintain an emf at its terminals; it stops when the dry cell is exhausted. Since current always flows in the *same* direction, the circuit is known as a **direct-current (d-c) circuit.**

Georg Simon Ohm discovered in 1827 that **the current flowing in such a d-c circuit is directly proportional to the applied voltage (emf) and inversely proportional to the resistance of the circuit.** Putting this statement, known as **Ohm's Law,** into mathematical form, we obtain

$$\text{CURRENT} = \frac{\text{EMF (VOLTAGE)}}{\text{RESISTANCE}}$$

or using symbols: $I \text{ (amperes)} = \dfrac{E \text{ (volts)}}{R \text{ (ohms)}}$

This equation gives the value of the current in a circuit when its resistance and applied voltage (emf) are known. Ohm's Law not only applies to a complete circuit, but also to any part of such a circuit, such as a single resistance. Thus, when the resistance (*R*) in ohms and the current (*I*) in amperes are known, the *voltage drop* (*E*) developed *across* the resistance is simply the *product* of the current and the resistance, or in equation form: VOLTAGE = CURRENT × RESISTANCE or E (volts) = I (amps) × R (ohms) (Note that we have used the symbol *E* here to represent a voltage *drop,* though it is used more correctly for an *emf.* The symbol *V*

is frequently used for a voltage drop or potential difference). The relation $E = I \times R$ may also be used to give the applied voltage (emf) acting in a **complete circuit,** if the current through the circuit and its total resistance are known.

If the emf (voltage) acting in a circuit and the current are known, the total resistance of the circuit, by Ohm's Law, is the **applied voltage divided by the current.** Putting this in mathematical form

$$\text{RESISTANCE} = \frac{\text{VOLTAGE}}{\text{CURRENT}}$$

or $R \text{ (ohms)} = \dfrac{E \text{ (volts)}}{I \text{ (amps)}}$

You can use this latter equation also to compute the resistance value of a **single resistor** in a circuit, by dividing the voltage drop developed across it by the current flowing through it. All three forms of Ohm's Law are used constantly in all types of electrical work. A few examples will help to clarify their use.

EXAMPLE 1: A resistor of 50 ohms is connected to a battery with an emf of 12 volts. What is the current through the resistor?

Solution: By Ohm's Law,

$$\text{the current } I = \frac{E}{R} = \frac{12}{50} = 0.24 \text{ ampere.}$$

EXAMPLE 2: A radio tube requires 0.15 ampere current to heat its filaments. What voltage must be applied to the filaments if their resistance is 42 ohms (when lit)?

Solution: The applied voltage $E = I \times R = 0.15 \times 42 = 6.3$ volts.

EXAMPLE 3: An ampere-meter (ammeter) inserted into a circuit reads a current of 5 amperes, when an emf of 170 volts is applied to the circuit. What is the total resistance of the circuit?

Solution: The circuit resistance

$$R = \frac{E}{I} = \frac{170}{5} = 34 \text{ ohms.}$$

Let us perform a simple experiment to confirm Ohm's Law and obtain an idea of the relationships involved:

EXPERIMENT 14: Obtain five small 1.5 volt flashlight batteries, five 10-ohm resistors (rated at 5 watts), an inexpensive 0-10 V voltmeter, a 0-1 A ammeter (or instead of two meters, a simple multimeter), and some connecting wire. Connect two of the 10-ohm resistors in series and hook the combination in *series* with the ammeter and one of the flashlight batteries, as shown in Fig. 55. Measure the exact emf of the cell by connecting the voltmeter *across it*, as shown. For an emf of 1.5 volts and a resistance of 20 ohms, the ammeter should indicate a current of 0.075 ampere (or 75 ma). Record this current on a piece of paper. (Note that we have neglected the resistance of the connecting wires and that of the ammeter, which should be very small compared to that of the 20-ohm lumped resistance.)

Now repeat the experiment by using two flashlight cells in series to obtain an emf of roughly 3 volts. Measure the exact emf by placing the voltmeter across both cells. The current indicated on the ammeter should have doubled—equal about 0.15 ampere. Record this current. Repeat the experiment three more times by connecting, in turn, three, four, and five flashlight cells in series to obtain emf's of about 4.5, 6, and 7.5 volts, respectively. Record the current in each case, which should go up in equal steps to (roughly) 0.225 amp., 0.3 amp., and 0.375 ampere, respectively. Plot a graph of current (I) against applied voltage (E) using the recorded data. This graph should be a *straight line,* as shown in Fig. 55, verifying that the **current in a d-c circuit is directly proportional to the applied emf (voltage).**

Let us now repeat the experiment, *keeping the applied emf constant* and *changing the resistance* instead, to obtain the variation of current with resistance. Connect four flashlight batteries in *series* to obtain a constant emf of about 6 volts. Measure the exact value by placing the voltmeter across the four cells. Connect this emf in series with the ammeter and with *one* 10-ohm resistor. (See Fig. 56.) The ammeter should read about 0.6 ampere. Record this value.

Repeat the experiment for resistance values of 20 ohms, 30 ohms, 40 ohms, and 50 ohms, by connecting, in turn, two, three, four and five of the 10-ohm resistors in series. Measure and record the ammeter current for each of these resistance values. The current should *decrease* from the original value of 0.6 ampere to (roughly) 0.3 amp. for 20 ohms, 0.2 amp. for 30 ohms, 0.15 amp. for 40 ohms, and finally about 0.12 amp. for 50 ohms. Now plot a graph of *current against resistance,* using the data obtained in the experiment. The graph should be a curved (hyperbolic) line, with the current *decreasing* with increasing resistance. The hyperbolic curve indicates that the current varies *inversely* with the re-

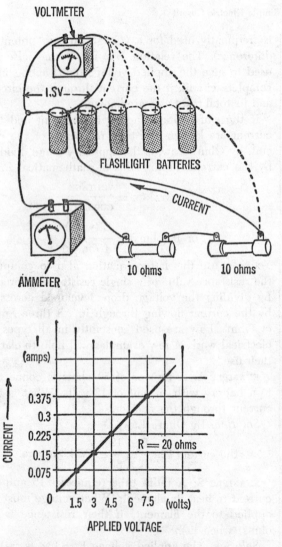

Fig. 55. Verifying that Current is Proportional to Voltage (Ohm's Law)

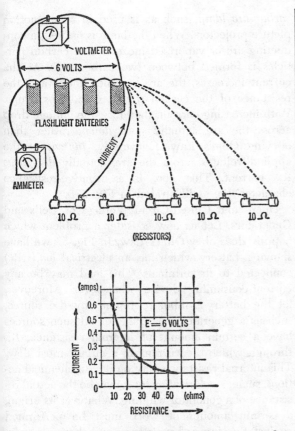

Fig. 56. Verifying that Current is Inversely Proportional to Resistance (Ohm's Law)

sistance, if the applied voltage is kept constant. You can verify that an inverse relation results in a hyperbola, by plotting the quotient of dividing a fixed number by increasingly larger numbers against these numbers. (For example, plot 1, 2, 3, 4, 5, . . . etc. against 1, ½, ⅓, ¼, ⅕.)

Where Ohm's Law does not Apply. Ohm's Law is *not* a universal law, like gravitation, but an experimental fact that holds for certain types of conductors. It does *not* apply to everything, though most electricians think it does. Our last experiment demonstrated an important fact; namely, that the plot of current versus voltage in a simple d-c circuit is a *straight line.* (See Fig. 55.) Ohm's Law applies wherever there is such a **linear relationship between voltage and current.** This will be the case as long as the **resistance** of a conductor or a circuit remains *constant,* regardless of the value of the current. Pure metals and metallic alloys have an essentially constant resistance, neglecting the small change in resistance due to the *heating* of the conductor when a current flows through it. Metals and alloys, therefore, are known as **linear conductors,**

and they are the only ones that obey Ohm's Law. Lest you despair, most d-c circuits are made up of such linear conductors and, hence, obey Ohm's Law.

Whenever the resistance of a device does *not* remain constant, the relationship between voltage and current will be **non-linear** (i.e., a curve) and Ohm's Law does *not* apply. Roughly, such a non-linear voltage-current relation exists in **semiconductors, electrolytes** and **ionized gases.** As a matter of fact, even the ordinary incandescent lamp is a *non-linear* conductor. The "hot" resistance of such a bulb, when it is brilliantly lit, may be some 15 to 20 times the "cold" resistance of the bulb, when no current flows through it. As a result, if you start increasing the applied voltage across such a lamp from zero to full voltage, the resistance of the bulb will go up almost as fast as the applied voltage and, hence, the current $(I = E/R)$ will remain practically *constant.* This **constant-current characteristic** is quite useful for regulating the amount of current flow through a circuit and lamps are used frequently for this purpose. Note that you can use Ohm's Law even for an incandescent lamp, provided you know two out of three quantities at all times. Thus, you can calculate the current through the lamp by Ohm's Law $(I = E/R)$ provided you know the voltage and the resistance *at the particular current.* Since the resistance is *not* independent of the current, this means that you have to measure or calculate the resistance in some *independent* way for each voltage or current value. You cannot *predict* the magnitude of the current for a particular voltage by Ohm's Law, since the resistance is *not a constant.*

Vacuum tubes, used in radios and TV sets, are an interesting *hybrid* of a linear and a non-linear device, exhibiting a little of both. As a voltage is applied across the tube and a current starts to flow, the internal resistance of the tube, which is initially high, drops rapidly to a fairly low value and then remains essentially constant. As a consequence, the current through the tube increases slowly at first with increasing voltage, then more rapidly as the resistance drops, and eventually increases almost linearly, as the resistance stabilizes. A typical voltage-current characteristic for a vacuum tube is shown in Fig. 57a. You can use Ohm's Law in the **linear (upper) portion** of the curve.

Certain semiconductors and carbon actually have a *negative* temperature-resistance characteristic; that is, their resistance *drops* as the temperature

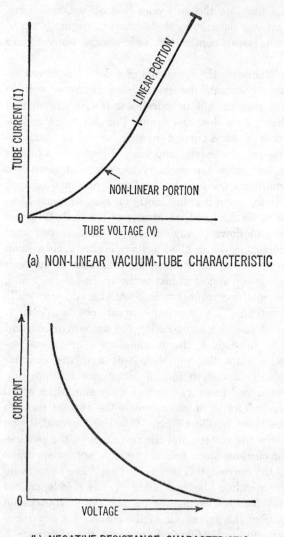

(a) NON-LINEAR VACUUM-TUBE CHARACTERISTIC

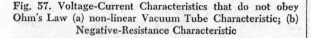

(b) NEGATIVE-RESISTANCE CHARACTERISTIC

Fig. 57. Voltage-Current Characteristics that do not obey Ohm's Law (a) non-linear Vacuum Tube Characteristic; (b) Negative-Resistance Characteristic

goes up. For instance, we have mentioned the thermistor, whose resistance may drop to as low as a millionth of its original value, when the temperature *increases* by some 50°C. Consider what happens when a current flows through a thermistor. The current will start to heat up the thermistor slightly and its resistance starts to drop rapidly. Since the resistance decreases much faster than the current can increase, the voltage drop across the thermistor (which is the product of current times resistance) actually *decreases* for an increasing current. Such a **negative-resistance characteristic** is illustrated in Fig. 57b.

The same sort of thing goes on in an electric *carbon-arc lamp,* such as is used in some motion picture projectors. When the lamp is ignited, a conducting arc of vaporized incandescent carbon particles is formed between two carbon tips. As the current increases, the arc becomes hotter and the resistance of the carbon particles and tips drops. With increasing current, therefore, the voltage drop across the arc actually goes *down,* rather than obeying Ohm's Law. Conversely, *increasing* the applied voltage across the arc, actually *decreases* the current. This, too, is a negative-resistance characteristic, as illustrated in Fig. 57b.

Open- and Closed-Circuit Voltage of Cells and Generators. Let us now consider a problem which happily *does* obey Ohm's Law. In Fig. 58 we have shown a battery which has an electrical *load* (R_L) connected to its terminals. This load may be any current-consuming device whatsoever. Moreover, let the battery symbol represent *any* d-c source, such as a generator, for example. All such sources have a certain amount of **internal resistance, Ri,** through which the current in a circuit must flow. This internal resistance may be due to chemical actions, such as polarization, or it may be the actual resistance of a generator winding. Whatever its origin, a certain amount of work must be performed against this internal resistance and, hence, a portion of the emf of the source is wasted in overcoming it. Let us place a resistor, equal in value to the internal resistance (Ri) of the source, in *series* with the source of emf and put the whole thing in a box. The potential difference (voltage) that appears across the terminals of the box is the voltage actually applied to the external circuit, or load. We would like to know the value of this **terminal voltage** (symbol V), both when the circuit is open (switch open) and also when the circuit is closed (switch closed).

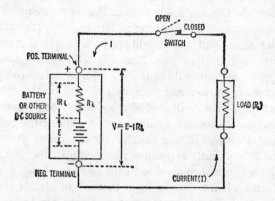

Fig. 58. Terminal Voltage of D-C Source

Let us apply Ohm's Law to solve this problem. By Ohm's Law the applied (terminal) voltage must equal the product of the current (I) through the circuit and the load resistance, R_L. Hence we write

$$V = I \times R_L \qquad (1)$$

But we have also established that the *emf of the source (E) must equal the sum of all the voltage drops* in the circuit. By Ohm's Law the voltage drop across the internal resistance (Ri) equals $I \times Ri$ and the voltage drop across the load is $I \times R_L$. Setting these voltage drops equal to the emf (E), we obtain

$$E = I\,Ri + I\,R_L \qquad (2)$$

Substituting $V = I\,R_L$ in equation (2); we write

$$E = I\,Ri + V \qquad (3)$$

and transposing, we obtain the result

$$\text{terminal voltage } V = E - I\,Ri \qquad (4)$$

which means that the terminal voltage in a closed circuit for a certain current flow (I) is simply the **emf minus the voltage drop across the internal resistance**. A little plain reasoning would have given us the same result without resort to mathematics. Moving in the direction of the electron current from the negative to the positive terminal, you encounter first a *rise* in potential (from $-E$ to $+E$) and then a fall in potential equal to $I \times R_i$. Subtracting the fall in potential from the rise to obtain the difference in potential between the terminals, V, you get the result above (i.e., $V = E - IR_i$). If the circuit is open (switch open), the current in equation (4) above is zero, and $V = E$; that is, the **open-circuit terminal voltage equals the emf of the source**.

EXAMPLE 1: What is the emf of a battery if its terminal voltage is 5.5 volts for a load current of 25 amperes, and the internal resistance of the battery is 0.02 ohm?

Solution: $E = V + I\,Ri = 5.5 + 25 \times 0.02 = 6$ volts.

EXAMPLE 2: Compute the internal resistance of a dry cell, which has an open-circuit voltage of 1.5 volts and a closed-circuit terminal voltage of 1.41 volts, when a current of 30 amperes is drawn.

Solution: Solving equation (4) for Ri, we obtain

$$Ri = \frac{E - V}{I} = \frac{1.5 - 1.41}{30} = \frac{0.09}{30} = 0.003 \text{ ohm.}$$

Practice Exercise No. 6

1. A wire carries a current of 15 amperes. How many coulombs pass a given point of the wire in 2 minutes?

2. Change 5 microvolts into volts; 15 ma into amps; and 2.5 megohms into ohms.

3. Define the international ampere and the absolute ampere.

4. Explain the energy concept of voltage and give the definition of volt.

5. State the definitions of the absolute ohm and the international ohm. Does Ohm's Law define the ohm?

6. If the cross-sectional area and length of a conductor are *doubled*, what happens to its resistance?

7. What should be the diameter of an 800-ft long copper wire to make its resistance 0.2 ohm?

8. What is the resistance of 1000 ft of 0.15 in. thick copper wire?

9. A 500-ft length of stranded copper wire is made up of 17 strands, each 0.032 in. in diameter. What is its resistance?

10. If the resistance of 1000 ft of No. 10 copper wire is 1 ohm, what is the resistance of 1000 ft of No. 6 copper wire?

11. The "cold" resistance of a 30-watt bulb is 32.4 ohms at 20°C and its "hot" resistance is 470 ohms. If the temperature coefficient of filament is 0.005, at what temperature does it burn?

12. A carbon filament lamp draws a current of 0.45 amp when a voltage of 122 volts is applied. What is its resistance?

13. A 20-ohm resistor is connected to a 6-volt battery. What current does it draw?

14. A toaster constructed of a 55-ohm resistance wire requires 4 amperes for its operation. What voltage should be applied?

15. An electrical device draws currents of 0, 0.5, 1.1, 1.8, and 2.6 amperes as the applied voltage is increased from 0 to 50 volts in 10-volt steps. Does the device obey Ohm's Law?

16. A battery with an internal resistance 0.25 ohm and an emf of 6.4 volts delivers a current of 2 amps to a load. What is (a) the terminal voltage of the battery and (b) the load resistance?

17. A dry cell has an open circuit terminal voltage of 1.476 volts and a closed-circuit terminal voltage of 1.435 volts when delivering a current of 0.558 ampere to a load. Find the internal resistance of the dry cell.

SUMMARY

Electric current is the **rate of flow** of electric charge. Hence, current equals **charge per unit time.** ($I = Q/t$)

The **potential difference** between two points is **1 volt** if **1 joule** of energy is either expended or required in moving a charge of **1 coulomb** from one point to the other. The **sum of all potential** (voltage) **drops** around a circuit **equals the emf of the** source.

The **opposition** to electron flow in a substance is

called **resistance** and the **ease** with which electrons pass through the material is called **conductance**. Conductance is the **reciprocal** of resistance.

A conductor across which a potential drop of 1 volt exists, when a current of 1 ampere flows through it, has a **resistance of 1 ohm**.

Resistivity is the resistance of a wire specimen of unit length and unit cross section. If the length is expressed in **centimeters** (cms) and the cross-sectional area in square centimeter (cm²), the resistivity is in **ohm-centimeters**. If the length is expressed in feet and the cross section in **circular mils** (1 mil = 0.001 inch; circular mils = (mils)²), the resistivity is in **ohm-circular mil per foot**, or briefly, **ohms per mil-foot**.

The **resistance of a conductor is directly proportional to its length** and **inversely proportional to its cross-sectional area**. Resistivity is the proportionality constant. $(R = \rho \dfrac{L}{A})$

The resistance of pure metallic conductors **increases with temperature**. The amount of increase depends on the **temperature coefficient of the resistance**. **Semiconductors** exhibit a **negative temperature-resistance characteristic**; that is, their resistance **decreases with temperature**. This property is used in **thermistors**.

Ohm's Law states that the **current flow in a d-c circuit is directly proportional to the applied voltage** (emf) and **inversely proportional to the resistance** of the circuit. $(I = E/R)$

Three forms of Ohm's Law: $I = E/R$; $E = I \times R$; $R = E/I$.

Ohm's Law applies to conductors made of **pure metals or metallic alloys**; it does **not** apply to semiconductors, electrolytes and ionized gases, and whenever the resistance is *not* a constant. A linear voltage-current graph demonstrates Ohm's Law; a non-linear voltage-current characteristic shows that Ohm's Law does *not* hold.

The **closed-circuit terminal voltage** of a d-c source (battery, generator, etc.) equals its **open-circuit emf** (voltage) minus the voltage drop across its internal resistance $(V = E - I\, Ri)$.

CHAPTER SEVEN

DIRECT-CURRENT CIRCUITS

Whenever there is a current flow, there must be an electrical **circuit**; that is, an **unbroken electrical pathway from source to load and back to source.** If all circuits were as simple as those described in the last chapter, a single application of Ohm's Law would suffice to determine the current, and there would be no need for the present chapter. Unfortunately, most practical circuits—even the ones used in your home—are not that simple. Sometimes a current flows consecutively through many different appliances (or loads) before returning to its source, in what is known as a **series circuit.** More often, the current flowing from a source will divide up into many different branches to feed houses, apartments, and the electrical devices in them, before it becomes re-united and returns to the source. This type of **divided current** flow is called a **parallel circuit.** Many actual circuits are a combination of both types, termed **series-parallel** circuits, with the current dividing into various **parallel** branches, each of which may have a number of loads connected in **series.**

Now we shall look into some of the methods used for "solving" these more complicated circuits. When we speak of "solving" a circuit, we generally mean three things. First, we would like to reduce the resistances offered by all the loads in the circuit to a single equivalent or **total resistance,** which will give us the **total current** withdrawn from the source of emf by an application of Ohm's Law (i.e., $I_{tot} = \dfrac{E}{R_{tot}}$). Secondly, we would like to know all the individual currents flowing through the various devices (loads) and branches of the circuit, giving us the **current distribution.** Finally, we want to determine the **fall of potential** or **voltage drop** across each of the loads to ascertain the **voltage distribution** in the circuit. You will find that Ohm's Law is constantly used in all these calculations, whenever it applies. But in addition, we shall learn more powerful methods of solving circuits, which will give us the answers quicker and with less trouble.

SERIES CIRCUITS

The connections illustrated in Figs. 54, 55, and 56 are all **series circuits** because in each one of them the current flows in an **undivided, consecutive and continuous path** from the source of emf through the various loads and back to the source. You can easily check whether a given arrangement of connections is a series circuit by imagining yourself to be an electric charge that travels from the negative terminal of the voltage source, through the circuit, to the positive terminal of the source. If you can move through the entire circuit in a single, continuous path, you have a series circuit; if you can find a way to return to the source after you break a connection *anywhere* in the circuit, you do *not* have a series circuit. By the way, the fact that a *single* interruption can stop *all* current flow in a series circuit is its biggest disadvantage. A series circuit either *operates all the way or not at all.* You can easily imagine what would happen if all the electric outlets in a city were connected in series with the power plant. If a single electric bulb burned out in some back street the whole city would be darkened.

Resistors in Series. Fig. 59 illustrates a simple series-connected circuit, similar to that shown in Fig. 56 (Experiment 14). Four resistors are connected in series with a 50-volt battery and a switch, which permits opening and closing the circuit. The resistors may represent any load, such as lamps or the filaments of radio tubes. We have also inserted an ammeter, symbolized by the circled (A), in series with the switch. It does not matter *where* the ammeter is inserted into the circuit, as long as the current flows *through* it, since the current is everywhere the *same.* In addition, we have shown a voltmeter, symbolized by the circled (V), connected *across* the battery, to indicate its emf. This voltmeter may be moved anywhere in the circuit to determine the voltage drops developed in the resistors. Note that an ammeter is always inserted in series with the part of the circuit through which the current is to be determined. A voltmeter, in contrast, is always connected *across* or in parallel with the part of the circuit across which the potential difference (voltage) is to be determined.

When we close the switch in the circuit of Fig. 59, a current flows in turn through each of the four resistors, the ammeter, the switch, and the bat-

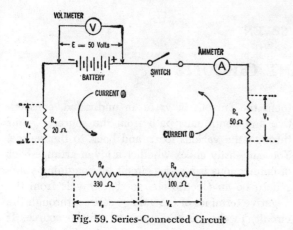

Fig. 59. Series-Connected Circuit

tery. Let us find the *value* of this current, not only for the circuit of Fig. 59, but in general for *any* series-connected circuit. Since, by definition, the current flows in a single, continuous path, we know at the outset that the **current in a series circuit must everywhere be the same.** We also know the emf (*E*) of the battery or other source. The only missing link is the **total resistance, R_t,** of the circuit. If we knew its value, we could determine the total series current (*I*) by Ohm's Law, thus:

$$\text{Total Series Current } I = \frac{E}{R_t} \quad (1)$$

Now we know "intuitively" and have assumed in the past that the **total opposition** (resistance R_t) to the current is the *sum* of the individual oppositions offered by the resistors. Thus we could simply state that the **total resistance (R_t) of the series circuit is equal to the sum of the individual resistance values** (R_1, R_2, R_3 and R_4). But it would be nice if we could *prove* this statement. To do this let us use the basic concept that **energy must be conserved.** Hence, the energy *expended* by the current in moving from the point of *high potential* (negative terminal) to the point of *low potential* (positive terminal) must *equal* the *energy* (or voltage) *supplied* by the source of emf (*E*). Equivalently, as we have stated before, *the sum of the individual voltage drops in the circuit must equal the emf* (E) *of the source.* Using the letter *V* to represent voltage drops, we can express this statement mathematically, as follows

$$E = V_1 + V_2 + V_3 + V_4 \quad (2)$$

where V_1 stands for the voltage drop across R_1, V_2 for that across R_2, and so on. By Ohm's Law, the voltage drop across each of the resistors is the product of the current (*I*) and the individual resistance (*R*). Since the current is everywhere the same, but

the resistance values differ, it is evident that the voltage drops in a series circuit can all be different. This is another important fact to remember. Applying it to the voltage drops across the resistors in the circuit of Fig. 59, we obtain

$$V_1 = I R_1; \ V_2 = I R_2; \ V_3 = I R_3; \text{ and } V_4 = I R_4 \quad (3)$$

We also know that the applied voltage (*E*) must equal the product of the current and the total resistance (R_t); that is, $E = I R_t$.

Substituting these relations in equation (2):

$$E = V_1 + V_2 + V_3 + V_4$$
$$I R_t = I R_1 + I R_2 + I R_3 + I R_4$$

factoring: $I R_t = I (R_1 + R_2 + R_3 + R_4)$

Dividing by "I": $R_t = R_1 + R_2 + R_3 + R_4 \quad (4)$

In general, therefore, for *any* number of series-connected resistors, R_1, R_2, R_3, R_4, ... etc., the total resistance

$$R_t = R_1 + R_2 + R_3 + R_4 + \ldots \quad (5)$$

where the dots represent any number of additional resistors used.

Let us summarize the relations we have just derived for a series circuit:

1. The **current** in a series circuit is everywhere the same.

2. The **voltage drops** may all be **different,** depending on the value of each resistance, but the sum of the voltage drops must add up to the **emf** (voltage) of the source.

3. The **total resistance** of a series circuit equals the sum of the individual resistances (or resistors).

EXAMPLE: Let us compute the total resistance and current in the circuit of Fig. 59. The total resistance $R_t = 50 + 100 + 330 + 20 = 500$ ohms. Hence, by Ohm's Law, the current $I = \dfrac{E}{R_t} = \dfrac{50}{500} = 0.1$ ampere.

Lamps or Tubes in Series. The relations we have worked out for resistors in series hold for any type of load, whether it be lamps, radio tubes, or anything else. Lamps are rarely connected in series, except in special cases, such as miniature Christmas tree lights, where each bulb has too low a voltage rating to be connected directly across the line voltage. If you have ever been annoyed by the burning out of one of these miniature bulbs and saw the whole string go out, you'll know why the series-connection of lamps is not generally in favor. Moreover, it is not at all easy to locate the defective lamp, since they all go out at the same time.

The series connection is frequently used for radio tubes. The filaments of vacuum tubes operate at a low voltage (1 to 50 volts) and, hence, cannot be connected *directly* across the 120-volt line. In inexpensive radio receivers, where a step-down filament transformer is not feasible, the filaments of all the tubes are connected in series and the entire string is connected to the power outlet. This can always be done, *provided* all the tubes are rated to operate at the same **current** and the **voltage ratings** of the individual tubes **add up to the line voltage** (usually 115 to 120 volts). If the voltages do not add up to the line voltage, an additional **ballast resistor** must be connected in **series** with the tubes to take up the **excess** line voltage.

EXAMPLE: As an example, let us compute the ballast resistance required for the filament circuit illustrated in Fig. 60. Here four tubes, each rated at 12.5 volts filament voltage, are connected in series with a 25-volt tube and the ballast resistor. All the tubes are rated to operate with a current of 0.15 ampere. The line voltage is 120 volts.

Solution: The voltage drops across the five tubes add up to $4 \times 12.5 + 25 = 75$ volts. The ballast resistor must develop a voltage drop equal to the excess, or the *difference* between the line voltage and the tube voltage drops. Hence, the voltage drop across ballast resistor $= 120 - 75 = 45$ volts. By Ohm's Law, the resistance of the ballast $= \dfrac{E}{I} = \dfrac{45}{0.15}$ $= 300\ ohms$.

Simple Voltage Divider (Potentiometer). You have already met a type of variable resistor, called a **potentiometer.** Let us see how a potentiometer divides the voltage applied to it **in proportion to** the resistance included between its movable contact and one of the fixed contacts. As shown in Fig. 61, the **input voltage** (E) from some source is applied between the two fixed ends (A and B) of the resistance winding (R) and the **output voltage** (V) is taken between the movable contact and the lower end.

Obviously, if we move the slider of the potentiometer all the way up to the top (point A), we shall tap off the full input voltage, E, and if we move it all the way to the bottom (point B), we shall get no output voltage at all. What is the output voltage

at some intermediate point (C)? Well, the current (I) in this simple series circuits is E/R. Hence, the voltage drop V across the portion of resistance, R', included between points C and B, is $I \times R'$. Substituting for I, we obtain $V = I\ R' = \dfrac{E}{R} \times R' = E \times$ $\dfrac{R'}{R}$, or the output voltage is the **input voltage times** the ratio of the two resistances.

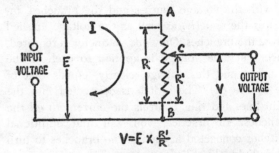

$$V = E \times \frac{R'}{R}$$

Fig. 61. Action of a Potentiometer-Voltage Divider

PARALLEL CIRCUITS

Most electrical circuits you will encounter in practice are parallel circuits. House and apartment wiring consists of a multiplicity of parallel connections and current paths, all fed by the same source of voltage. The parallel circuit is basically different from the series type, inasmuch as the current divides into a number of separate, independent branches. Each of these branches may have a different resistance (load) and, hence, the value of the current in each branch may be different. If one of the loads burns out or is disconnected, the remainder of the circuit continues to function, which is a great advantage over the series circuit. Moreover, since all branches operate on the same voltage source, a single power source of the proper voltage and power rating can supply the currents to all the parallel branches. The check for a parallel circuit is simple: If you can trace **more** than one path for the current to flow through the circuit, you have a parallel circuit.

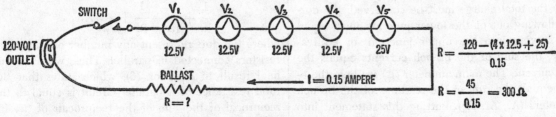

Fig. 60. Series-Connected Filament Circuit of a Radio Receiver

Resistors in Parallel. Let us now solve an actual parallel circuit, consisting of three resistors (R_1, R_2 and R_3) connected in parallel across a voltage source (E). As before, the resistors may represent any electrical appliance or load of a certain resistance value. The arrangement is illustrated in Fig. 62. We have inserted ammeters (A_1, A_2, and A_3) into each of the three branches to indicate the individual currents, and also one (A) into the **main line** to indicate the total current (I). A voltmeter (V), connected in parallel with the voltage sources and the branches, indicates the emf (E) as well as the voltage applied across the branches. Only *one* voltmeter is required, since the same voltage is applied to each of the branches and this voltage, clearly, equals that of the source (E). The switch is connected into the main line, and thus controls the current to *all* the branches. In practice, you will find additional switches connected in each of the branches to turn the individual appliances on or off.

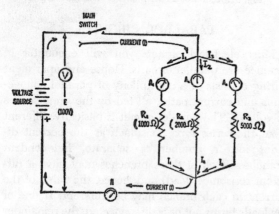

Fig. 62. Parallel-Connected Circuit

When the main switch is closed, a current (I) flows from the source of emf (E) in the direction indicated by the arrows to a common **upper junction** point of the three resistors (R_1, R_2 and R_3). At this point the current divides into three branch currents, I_1, I_2, and I_3, respectively. After flowing through the individual resistors, the branch currents again combine at the **lower junction** point. Since the total charge must be conserved, the current flowing out of the lower junction must equal that flowing into the upper junction, or equivalently, the **sum of the branch currents equals the total current.** The main ammeter (A) reading, therefore, equals the *sum* of the readings of the branch ammeters (A_1, A_2, A_3). Putting this statement into mathematical form, we obtain

Total current $I = I_1 + I_2 + I_3 =$ sum of branch currents (1)

The parallel circuit thus acts as a **current divider,** in contrast to the series circuit, which is a **voltage divider,** as we have seen.

By Ohm's Law, the voltage drop across each branch resistor is the product of the branch current and the branch resistance, and this product must equal the emf of the source (E). The total current (I) is, of course, the emf (E) divided by the total resistance (R_t), or $\dfrac{E}{R_t}$. Moreover, since the total current is *greater* than the current through any branch, the total resistance ($R_t = \dfrac{E}{I}$) must be less than the lowest value of any branch resistance. Putting these considerations into equation form, we may write

$$I = \frac{E}{R_t} \; ; \; I_1 = \frac{E}{R_1} \; ; \; I_2 = \frac{E}{R_2} \; ; \; I_3 = \frac{E}{R_3} \; : \quad (2)$$

Substituting for the currents in eq. (1):

$$I = I_1 + I_2 + I_3$$
$$\frac{E}{R_t} = \frac{E}{R_1} + \frac{E}{R_2} + \frac{E}{R_3} \quad (3)$$

and dividing both sides by E, we obtain

$$\frac{1}{R_t} = \frac{1}{R_1} + \frac{1}{R_2} + \frac{1}{R_3} \quad (4)$$

Equation (4) states that the reciprocal of the total resistance in a parallel circuit equals the sum of the reciprocals of the individual branch resistances. Remembering that conductance is the reciprocal of resistance ($G = 1/R$), equation (4) also means that the **total conductance of a parallel circuit is the sum of the individual conductances.** Expressed mathematically,

$$G = G_1 + G_2 + G_3 \quad (mhos) \quad (5)$$

where $G = 1/R_t$; $G_1 = 1/R_1$; $G_2 = 1/R_2$; and $G_3 = 1/R_3$.

Equation (4) may be solved for the total resistance, R_t, giving:

$$R_t = \frac{1}{\dfrac{1}{R_1} + \dfrac{1}{R_2} + \dfrac{1}{R_3} + \ldots \ldots} \quad (6)$$

where the dots represent any number of additional resistors connected in parallel. Thus, we have our final result in equation (6), which states that the **total resistance of a parallel circuit is equal to the reciprocal of the sum of the reciprocals of the individual branch resistances.**

You do not always have to use this cumbersome formula (eq. 6), since the parallel circuit is often very simple. For example, if all the branch resistors are *identical*, the total resistance is simply the resistance of one branch divided by the number of branches, as we have already seen.

EXAMPLE 1: What is the total resistance of five 100-ohm resistors connected in parallel?

Solution: The total resistance $R_t = \dfrac{100}{5} = 20$ ohms.

As another special case, which occurs frequently in practice, consider the total resistance of *two* resistors connected in parallel. Equation (4) then becomes

$$\frac{1}{R_t} = \frac{1}{R_1} + \frac{1}{R_2} = \frac{R_1 + R_2}{R_1 R_2}$$

and hence, $R_t = \dfrac{R_1 R_2}{R_1 + R_2}$ \hfill (7)

In words, the **total resistance of two parallel branches is the product of the two resistances divided by their sum.**

EXAMPLE 2: What is the total equivalent resistance of a 20-ohm and a 80-ohm resistor connected in parallel?

Solution:

$$R_t = \frac{R_1 R_2}{R_1 + R_2} = \frac{20 \times 80}{20 + 80} = \frac{1600}{100} = 16 \text{ ohms.}$$

Let us summarize the relations we have just developed for a parallel circuit:

1. The **voltage drop** across each branch of a parallel circuit is the **same** and is **equal to the voltage of the source.**

2. The **total current** flowing into and out of the junction points of the branches equals the **sum of the branch currents.**

3. The **total conductance** of a parallel circuit equals the **sum** of the individual **branch conductances.**

4. The **total (equivalent) resistance** of a parallel circuit is equal to the **reciprocal of the sum of the reciprocals of the individual branch resistances.**

5. The **total resistance of identical parallel resistors** is the value of **one resistor divided by the number of branch resistors.**

6. The **total resistance of two resistors in parallel** is the product of the two resistance values divided by their sum.

EXAMPLE 3: As another example let us work out the total resistance, the total current and the branch

currents for the circuit of Fig. 62. By equation (6), the total resistance of the circuit,

$$R_t = \frac{1}{1/1000 + 1/2000 + 1/5000} = \frac{1}{\dfrac{10 + 5 + 2}{10,000}} =$$

$$\frac{10,000}{17} = 588 \text{ ohms}$$

Since we now have the total resistance, which is the *equivalent* of the three branch resistances, we can replace the circuit of Fig. 62 by a simpler, equivalent circuit, having only a single 588-ohm resistor. The total current for the equivalent circuit

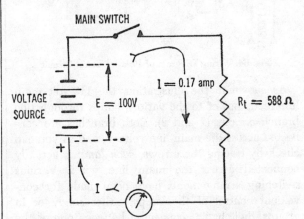

Fig. 63. Equivalent Circuit for that Shown in Fig. 62

shown in Fig. 63, is simply $I = \dfrac{E}{R_t} = \dfrac{100}{588} = 0.17$ ampere. To obtain the individual branch currents, we must go back to the original circuit of Fig. 62. Here, the current through R_1: $I_1 = \dfrac{E}{R_1} = \dfrac{100}{1000} = 0.1$ ampere

the current through R_2: $I_2 = \dfrac{E}{R_2} = \dfrac{100}{2000} = 0.05$ ampere

the current through R_3: $I_3 = \dfrac{E}{R_3} = \dfrac{100}{5000} = 0.02$ ampere

As a check let us add up the branch currents, which must result in the total current. Thus, the total current

$$I = I_1 + I_2 + I_3 = 0.1 + 0.05 + 0.02 = 0.17 \text{ ampere,}$$

or the same value we obtained before.

Lamps in Parallel. Fig. 64 illustrates an actual wiring diagram of a number of lamps connected in parallel. The diagram may at first appear complicated because of the different switch connections, but when you trace it out, it is just a combination of two simple parallel circuits.

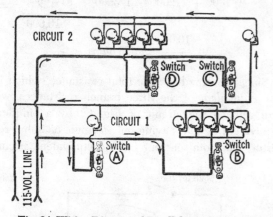

Fig. 64. **Wiring Diagram of Parallel Lamp Circuit**

As shown in the illustration, the 115-volt main line feeds power to the various lamps through *two* branch circuits (*1* and *2*). Both branches are connected across the main line voltage, and as you can check by tracing the arrows, *each lamp* is actually connected across the main line wires. Various switching arrangements have been made for convenient control of the lamps, either singly or in groups. Switch *A* is connected between one of the branch wires of circuit 1 and a single lamp. It therefore turns this lamp on or off. (The switch could have equally well been connected on the *other side* of the lamp.) Switch *B* is connected in series with a wire feeding five parallel lamps in circuit 1. This switch, therefore, turns all five bulbs either on or off, *as a group*. Switch *C* controls a single lamp in circuit 2 and switch *D* controls a group of five parallel lamps in circuit 2. Their functions are the same as those of circuit 1.

SERIES-PARALLEL CIRCUITS

As the name implies, series-parallel circuits are combinations of both series- and parallel-connected circuits and hence have the properties of both. No generalizations can be made about these circuits beyond that the series- and parallel portions must be solved separately by the methods we have developed. Ohm's Law is very inefficient to solve the more complicated series-parallel circuits and we shall learn better ways of dealing with them in the next section. The simpler types of series-parallel circuits yield to Ohm's Law, if you use it systematically. The general method is to **simplify the circuit step by step by replacing groups of series or parallel resistances by equivalent single resistances**, reducing it finally to a simple series circuit. It is best to find first the equivalent resistance of the parallel groups and then add to this the sum of the series-connected parts of the circuit. You will eventually end up with a single equivalent resistance (R_t) and a single source of emf (E); and hence you can determine the total current in the series-parallel circuit by Ohm's Law ($I = E/R_t$). By further, repeated applications of Ohm's Law you can then find the current in each branch resistance and the voltage drop across each. The following example will illustrate the method used. (See Fig. 65.)

EXAMPLE: Fig. 65 illustrates a series-parallel circuit consisting of a 100-volt source of emf (E) and five resistors, connected in series-parallel, with values as indicated in the figure. It is desired to find the total equivalent resistance (R_t), the total line current (I), the value of the current through each resistor, and the voltage drop across each.

Solution: The circuit of part I, Fig. 65, must first be simplified to a simple series circuit, as is indicated by the successive equivalent circuits, illustrated in parts II through IV. First combine resistors R_3 and R_4 into a single equivalent resistance, R_{3-4}. By the formula for two parallel resistors (eq. 7), we obtain

$$R_{3-4} = \frac{5 \times 20}{5 + 20} = \frac{100}{25} = 4 \text{ ohms.} \quad \text{(Part II)}$$

This equivalent 4-ohm resistor may be combined with the series-connected 16-ohm resistor (R_5) to yield an equivalent resistance of $4 + 16 = 20$ ohms. Now combine this 20-ohm resistor with the parallel-connected 80-ohm resistor, R_2, to obtain the equivalent resistance

$$R_{2,3,4,5} = \frac{20 \times 80}{20 + 80} = \frac{1600}{100} = 16 \text{ ohms.} \quad \text{(Part III)}$$

Finally, there remains only the 4-ohm resistor R_1, which is connected in series with the equivalent 16-ohm resistance ($R_{2,3,4,5}$). Combining these two resistors, we obtain

$$R_t = 16 + 4 = 20 \text{ ohms} \quad \text{(Part IV)}$$

The total current is therefore $I = \dfrac{E}{R_t} = \dfrac{100}{20} =$ *5 amperes*. Now let us determine the individual currents and voltage drops. The current through R_1 is the total line current, or *5 amperes*. Hence, the voltage drop across R_1 equals $5 \times 4 = $ *20 volts*. The

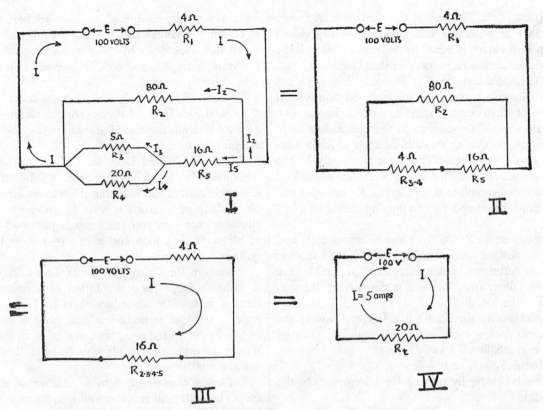

Fig. 65. Series-Parallel Circuit and its Equivalents

voltage drop across the entire combination of R_2, R_3, R_4 and R_5, or its equivalent resistance $R_{2,3,4,5}$ is the line current times the equivalent resistance, which is $5 \times 16 = 80$ volts. (It is also equal to the *difference* between the emf and the voltage drop across R_1, or again $100 - 20 = 80$ volts.)

The voltage drop across the 80-ohm resistor (R_2) is the same as that across the entire combination, or 80 volts. Hence the current through R_2 is $\frac{80}{80} = 1$ ampere. The current through R_5 is the voltage drop across $R_{2,3,4,5}$ divided by the resistance of the branch where R_5 is located. This resistance is $4 + 16 = 20$ ohms, as indicated in Part II. Hence the current through R_5 is $\frac{80}{20} = 4$ amperes (it is also the difference of the total current minus that through R_2, or $5 - 1 = 4$ amps). The voltage drop across parallel resistors R_3 and R_4 is the branch current times the equivalent resistance R_{3-4}, or $4 \times 4 = 16$ volts. The voltage drop across R_5 is this same branch current times R_5, or $4 \times 16 = 64$ volts. The two voltage drops add up to 80 volts, as expected.

Finally, the current through R_3 equals the branch voltage (16 volts) divided by R_3, or $\frac{16}{5} = 3.2$ *amperes*. The current through R_4 is correspondingly $\frac{16}{20} = 0.8$ *ampere*. This completes the solution of the entire circuit.

KIRCHHOFF'S LAWS

The example we have just considered indicates how involved the solution of a relatively simple series-parallel circuit may become, if carried out by Ohm's Law. More intricate circuits cannot readily be solved by the method just illustrated. Fortunately, two generalizations first suggested by the German physicist GUSTAV ROBERT KIRCHHOFF (1824-1887) have been found extremely useful for solving these more complicated circuits. Kirchhoff's observations are valid for *any* electrical circuit and they are known as **Kirchhoff's Laws**. In brief, Kirchhoff's Laws state:

First Law: **The sum of the currents flowing into a junction of an electric circuit is equal to the sum of the currents flowing out of the junction.** In other words, as much current flows *away* from a point as flows *towards* it.

Second Law: **The sum of the electromotive forces (battery or generator voltages) around any closed loop of a circuit is equal to the sum of the voltage drops across the resistances in that loop.**

Kirchhoff's first "law" is obvious; if it were not true more electric charges might be flowing *toward* a point than *away* from it, and, consequently, charges would *accumulate* at that point. But there is no reason that this should happen at some arbitrary point in a circuit. The law is thus equivalent to the statement that (for steady currents) **electricity does not accumulate at a junction.** An example will illustrate the simplicity of applying Kirchhoff's first law.

EXAMPLE: In Fig. 66 two branch currents (I_1 and I_2) are flowing towards a junction (L) of an electric circuit and three branch currents (I_3, I_4 and I_5) are flowing away from it. What is the current distribution at the junction?

Solution: By Kirchhoff's first law, the sum of the currents flowing into the junction equals the sum of the currents flowing away from it.

Hence: $I_1 + I_2 = I_3 + I_4 + I_5$
Or equivalently, by transposing all currents to the left side

$$I_1 + I_2 - I_3 - I_4 - I_5 = 0$$

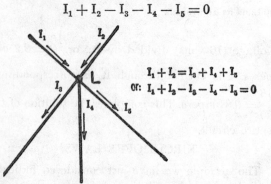

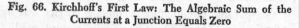

Fig. 66. Kirchhoff's First Law: The Algebraic Sum of the Currents at a Junction Equals Zero

This example suggests that Kirchhoff's first law may be further simplified. By assigning a *plus* (+) *sign* to all currents flowing *towards* a junction and a *minus* (−) *sign* to the currents flowing *away* from the junction, you can rephrase the first law as follows: **The algebraic sum of the currents at a junction is zero.** Putting this statement into mathematical form, we write concisely

Kirchhoff's First Law: Sum $I = 0$.
Kirchhoff's second law is not new to us; we have used it several times before. You will recall that it is based on the concept of **conservation of energy;** specifically, the energy delivered to a circuit must

equal the energy consumed, or the work performed in creating a voltage rise at the source of emf must equal the work done by the current in generating the voltage drops in a circuit. Expressed as an equation, the law states

Sum of Emf's = Sum of Voltage drops
or Sum E = Sum IR drops (around *closed* loop)
and by transposing the IR drops to the left side, we may write

Kirchhoff's Second Law: Sum E − Sum IR = 0
This may also be stated in words as follows: **The algebraic sum of the potential differences around a closed loop of a circuit is zero.** In order to obtain the *algebraic* sum you must assign *plus* (+) *signs* to all *emf's* in a loop and *minus* (−) *signs* to all *voltage* (IR) *drops.*

Procedure for Using Kirchhoff's Laws. In using Kirchhoff's Laws for the solution of actual problems, a systematic procedure must be followed in order not to get enmeshed in a number of confusing and interrelated equations, and to solve the circuit in the quickest possible way. In brief, these are the essentials:

1. Divide the circuit into a number of *closed loops,* including all resistors and sources of emf.

2. Assign a *direction* of *current* (electron) *flow* around each of the loops. If an emf (battery or generator) is present in the loop, choose the direction of *electron flow,* from minus (−) to plus (+). If no emf or several emf's are present, *assume* a current direction arbitrarily, either clockwise or counterclockwise. (NOTE: If you assume the wrong current direction, the value of the current will come out negative (−), but its *magnitude will not be affected.*) Then assign a *plus* (+) *sign* to those emf's that *tend* to produce currents in the *chosen* direction and assign a *minus* (−) *sign* to all emf's and currents flowing *opposite* to the chosen direction.

3. Using Kirchhoff's first law (Sum I = 0) write as many *independent* current equations as possible at various junction points. Currents flowing *into* a junction are *plus* (+), those flowing *out* of the junction are *minus* (−).

4. Using Kirchhoff's second law (Sum E = sum IR drops) write as many *independent* voltage equations around *closed loops,* as there are loops. The total number of independent current and voltage equations must **equal the number of unknown currents.** (Independent equations will *not* reduce to *identical* forms by algebraic substitution.)

5. Solve the resulting *simultaneous algebraic* equation for any or all desired currents.

The following examples will clarify the procedure used.

EXAMPLE 1: Fig. 67 shows a circuit containing two emf's and three resistors. It is desired to find the current *through* and the voltage drop *across* each resistor, using Kirchhoff's laws.

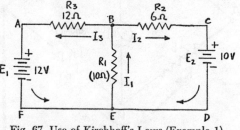

Fig. 67. Use of Kirchhoff's Laws (Example 1)

Solution: Let us choose loops ABEF and BCDE to solve the circuit. The current flow in each loop is chosen in the direction from the negative to the positive terminal of the emf. Let I_1 be the current through the 10-ohm resistor (R_1), I_2 the current through the 6-ohm resistor (R_2) and I_3 the current through the 12-ohm resistor (R_3).

By Kirchhoff's first law, at junction *B*:

$$I_1 = I_2 + I_3 \text{ or } I_1 - I_2 - I_3 = 0 \qquad (1)$$

If you try to write another current equation, at junction *E*:

$$I_2 + I_3 = I_1 \text{ or again } I_1 - I_2 - I_3 = 0$$

you will find that the latter equation is the same as eq. (1) and hence *not* independent from it.

By Kirchhoff's second law, the voltage drops around loop ABEF:

$$10\,I_1 + 12\,I_3 = 12 \text{ V} \qquad (2)$$

and the voltage drops around loop BCDE:

$$10\,I_1 + 6\,I_2 = 10 \text{ V} \qquad (3)$$

Substituting for $I_3 = I_1 - I_2$ in eq. (2)

$10\,I_1 + 12\,(I_1 - I_2) = 12$ V and hence

$$22\,I_1 - 12\,I_2 = 12 \text{ V} \qquad (4)$$

Dividing eq. (4) by 2: $11\,I_1 - 6\,I_2 = 6$ V $\qquad (5)$

Adding (3) to (5):

$$21\,I_1 = 16 \text{ V}$$

and hence $I_1 = \dfrac{16}{21} = 0.762$ *ampere*

from eq. (3):

$$I_2 = \frac{10 - 10\,I_1}{6} = \frac{10 - 7.62}{6} = 0.397 \text{ ampere}$$

and from eq. (1):

$$I_3 = I_1 - I_2 = 0.762 - 0.397 = 0.365 \text{ ampere}$$

Knowing the currents, we can now compute the voltage drops:

The voltage drop across the 10-ohm resistor (R_1)

$$I_1\,R_1 = 0.762 \times 10 = 7.62 \text{ volts}$$

The voltage drop across the 6-ohm resistor (R_2)

$$I_2\,R_2 = 0.397 \times 6 = 2.38 \text{ volts}$$

And the voltage drop across the 12-ohm resistor (R_3)

$$I_3\,R_3 = 0.365 \times 12 = 4.38 \text{ volts}$$

EXAMPLE 2: Fig. 68 illustrates a *network* of resistors connected to a 100-volt source of emf. Find the current through and the voltage across each resistor, the total current and the total resistance of the network, using Kirchhoff's Laws.

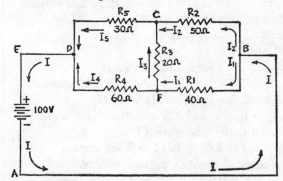

Fig. 68. Use of Kirchhoff's Laws (Example 2)

Solution: Let us choose loops ABFDE, ABFCDE, and ABCDE to solve the circuit. The current direction is from minus to plus, as indicated by the arrows. Assume that current I_3 through R_3 flows from F to C. Then, by Kirchhoff's first law, at junction *F*:

$$I_1 = I_3 + I_4 \text{ or } I_4 = I_1 - I_3 \qquad (1)$$

also, at junction *C*:

$$I_5 = I_2 + I_3 \qquad (2)$$

Using eq. (1) and (2) to write voltage drops around the loops, we obtain in loop ABFDE:

$$40\,I_1 + 60\,(I_1 - I_3) = 100 \text{ V}$$

or $\qquad 100\,I_1 - 60\,I_3 = 100 \qquad (3)$

in loop ABFCDE:

$$40\,I_1 + 20\,I_3 + 30\,(I_2 + I_3) = 100 \text{ V}$$

or $\qquad 40\,I_1 + 30\,I_2 + 50\,I_3 = 100 \qquad (4)$

and in loop ABCDE:

$$50\,I_2 + 30\,(I_2 + I_3) = 100 \text{ V}$$

or $\qquad 80\,I_2 + 30\,I_3 = 100 \qquad (5)$

Multiplying eq. (5) by 2

$$160\,I_2 + 60\,I_3 = 200 \qquad (6)$$

Adding eq. (3) to eq. (6)

$$100\,I_1 + 160\,I_2 = 300$$

or $\qquad I_1 = \dfrac{300 - 160\,I_2}{100} = 3 - 1.6\,I_2 \qquad (7)$

Solving eq. (5) for I_3

$$30\,I_3 = 100 - 80\,I_2$$

and hence $\qquad I_3 = 3.33 - 2.667\,I_2 \qquad (8)$

Substituting eqs. (7) and (8) in eq. (4)

$$40\,(3 - 1.6\,I_2) + 30\,I_2 + 50\,(3.33 - 2.667\,I_2) = 100$$

Multiplying out

$120 - 64 I_2 + 30 I_2 + 166.5 - 133.3 I_2 = 100$
Transposing
$120 + 166.5 - 100 = 133.3 I_2 - 30 I_2 + 64 I_2$
Simplifying
$$186.5 = 167.3 I_2$$

and hence $I_2 = \dfrac{186.5}{167.3} = 1.115\ ampere$

Substituting for I_2 in eq. (7)
$I_1 = 3 - 1.6\ (1.115) = 3 - 1.78 = 1.22\ ampere$
Substituting for I_2 in eq. (8)
$I_3 = 3.33 - 2.667\ (1.115) = 3.33 - 2.97 =$
$0.36\ ampere$

Substituting for I_1 and I_3 in eq. (1)
$I_4 = I_1 - I_3 = 1.22 - 0.36 = 0.86\ ampere$
Substituting for I_2 and I_3 in eq. (2)
$I_5 = I_2 + I_3 = 1.115 + 0.36 = 1.475\ ampere$
At point B, the total current $I = I_1 + I_2$
or $I = 1.22 + 1.115 = 2.335\ ampere$
As a check, at point D, the total current $I = I_4 + I_5$
or $I = 0.86 + 1.475 = 2.335\ ampere$,
or the same as above.

The total resistance, $R_t = \dfrac{E}{I} = \dfrac{100}{2.335} = 42.8\ ohms$

The voltage (IR) drops across resistors R_1 through R_5 are:

across R_1 : $I_1 R_1 = 1.22 \times 40 = 48.8\ volts$
across R_2 : $I_2 R_2 = 1.115 \times 50 = 55.75\ volts$
across R_3 : $I_3 R_3 = 0.36 \times 20 = 7.2\ \ \ volts$
across R_4 : $I_4 R_4 = 0.86 \times 60 = 51.6\ \ volts$
and across R_5 : $I_5 R_5 = 1.475 \times 30 = 44.25\ volts$
This completes the solution of the network shown in Fig. 68.

The Wheatstone Bridge. The network illustrated in Fig. 68 is actually a **bridge circuit,** as you can see by comparing it with Fig. 69, where we have redrawn the original circuit. The form of the circuit shown in Fig. 69 is known as the **Wheatstone Bridge** in honor of the British physicist SIR CHARLES WHEATSTONE (1802-1875). The Wheatstone Bridge is used for highly precise resistance measurements, by comparing an unknown resistance, R_x, with a known *standard* resistance, R_s.

As illustrated in Fig. 69, four resistors R_a, R_b, R_s and R_x are arranged in the form of a parallelogram and are connected to a battery E at two junction points A and C. A galvanometer with an internal resistance R_g bridges the other two junction points B and D. One of the resistors, R_x, is unknown in value, while the other three are known and adjustable. In practice, resistors R_a and R_b are given suit-

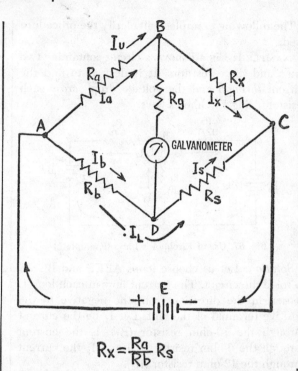

$$R_x = \frac{R_a}{R_b}\, R_s$$

Fig. 69. Schematic Circuit Diagram of Wheatstone Bridge

able fixed values and the standard resistance R_s is adjusted in value, until the galvanometer indicates *zero current flow.* The bridge is then said to be **balanced.** For a balanced bridge, the unknown resistance, R_x, is easily determined from a knowledge of the values of R_a, R_b and R_s.

Let us derive the equation for a balanced Wheatstone Bridge; that is, when no current flows through the galvanometer (G). If no current is diverted through the galvanometer, the current through R_a must be the same as that through R_x, or $I_a = I_x$. Let us label this *upper branch current* I_u. Similarly, the current I_b through R_b must equal the current I_s through R_s, since no current flows out of the lower junction D. Call this lower branch current I_l. Moreover, in the absence of a current through G, the potential at point B must equal that at point D. It follows that the voltage drop across R_a must be the same as the voltage drop across R_b, and the voltage drop across R_x must equal that across R_s. Let us put these conclusions into mathematical form:

$$I_a = I_x = I_u \qquad (1)$$
$$I_b = I_s = I_l \qquad (2)$$
the voltage drops $I_u R_a = I_l R_b \qquad (3)$
and $I_u R_x = I_l R_s \qquad (4)$
Dividing eq. (3) by eq. (4), term by term

$$\frac{R_a}{R_x} = \frac{R_b}{R_s} \qquad (5)$$

Solving for R_x, we obtain

$$R_x = \frac{R_a}{R_b} R_s \qquad (6)$$

Since the value of R_a, R_b and R_s are known, we can determine the value of the unknown resistor R_x from equation (6).

Practice Exercise No. 7

1. State the distinguishing characteristics of a series circuit and give its advantages and disadvantages.

2. Find the voltage (emf) required to send a total current of 6 amps through three series-connected resistors of 4, 8, and 10 ohms, respectively. Also compute the voltage drop across each resistor.

3. A lamp has a resistance of 150 ohms and the connecting wires have a resistance of 10 ohms total. An emf of 120 volts is applied. Find the current in the line, the voltage across the lamp, and the voltage drop in the line.

4. Eight Christmas tree lamps, each having 30 ohms resistance, are connected in series to the 120-volt line. What is the current through the first lamp? The last lamp?

5. A 4-ohm resistor, a 6-ohm resistor, and one of unknown value are connected in series to a 120-volt source. If the voltage drop in the 6-ohm resistor is 60 volts, what is the value of the unknown resistor?

6. Distinguish between series and parallel circuits and state the advantages of the parallel connection.

7. Show that for two resistors, R_1 and R_2, connected in parallel the branch currents, I_1 and I_2, divide in accordance with the relation $\dfrac{I_1}{I_2} = \dfrac{R_2}{R_1}$

8. A circuit has three parallel branches with resistances of 20, 30, and 40 ohms, respectively. If a current of 3 amps is flowing through the 30-ohm branch, how much current flows in each of the other branches? What is the applied emf, the total circuit resistance, and the total current?

9. What resistance must be placed in parallel with a 16-ohm resistor to make the total resistance of the combination 12 ohms?

10. A 120-volt electric heating appliance has a 12-ohm and a 24-ohm coil. If the coils are connected in parallel, compute the current drawn by each coil and the total current in the appliance.

11. A 2-ohm and a 4-ohm resistor are connected in parallel and the combination is connected in series with a 5-ohm resistor and a 3-volt battery with an internal resistance of 0.8 ohm. What is the current in the 4-ohm resistor?

12. Seven 200-ohm resistors are arranged in two groups, one of four parallel resistors and one of three parallel resistors. Both groups are connected in series with each other and with a 220-volt power line. What is the voltage drop across each group?

13. Two 3-ohm coils are connected in series and one of them is shunted by a 6-ohm coil. What is the total resistance?

14. Solve the circuit shown in Fig. 65 (I) by means of Kirchhoff's Laws.

15. Show that when the battery and galvanometer in a *balanced* Wheatstone Bridge (Fig. 69) are interchanged, the bridge will remain balanced.

SUMMARY

In a **series circuit** the current flows in an undivided, consecutive and continuous path from the source of emf through the loads and back to the source.

The **current in a series circuit** is everywhere the same.

The **voltage drops in a series circuit** depend on the values of the series resistances; the *sum* of the voltage drops must equal the *emf* of the source.

The **total resistance of a series circuit** equals the *sum* of the individual resistances (or resistors).

In a **parallel circuit** the current divides into a number of separate, independent branches. The following relations hold:

1. The **voltage drop** across each branch of a parallel circuit is the *same* and **equal to the voltage of the source.**

2. The **total current** flowing into and out of the junction points of the branches **equals the sum of the branch currents.**

3. The **total conductance** equals the *sum* of the branch conductances.

4. The **total equivalent resistance** equals the reciprocal of the sum of the reciprocals of the individual branch resistances.

5. The **total resistance of identical** parallel resistors is the value of *one* resistor *divided by the number* of branch resistors.

6. The **total resistance of two resistors in parallel** is the *product* of the two resistance values *divided by their sum.*

Kirchhoff's First Law: The **sum of the currents** flowing *into* a junction of a circuit **equals the sum** of the currents *flowing out.* (The *algebraic* sum of the junction currents is zero; Sum I = 0)

Kirchhoff's Second Law: The **sum of the emf's** around any **closed loop** of a circuit *equals* the sum of the **voltage drops** across the loop resistances. (Sum E − Sum IR = 0)

CHAPTER EIGHT

ELECTRIC POWER AND HEAT

Your experience with household electric appliances indicates that heat is produced whenever an electric current flows through a resistance. You'll recall that this heat is caused by collisions between free electrons moving through the conductor and the relatively "fixed" atoms making up the crystal structure of the conductor. These collisions increase the kinetic or *thermal* energy of the atoms within the conductor and its temperature rises. The more current flows, the greater is the increase in the thermal energy of the conductor and, hence, the greater is the heat liberated. An experiment will confirm this.

EXPERIMENT 15: Obtain a few feet of No. 30 *nichrome* resistance wire, a pyrex, porcelain or other heat-resistant tube about ½-inch in diameter; and a large rheostat that will permit a current flow of from 1 to 10 amps. Wind about 2 feet of nichrome wire around the heat-resistant tube and connect the ends in series with the rheostat and the 120-volt power line, as illustrated in Fig. 70.

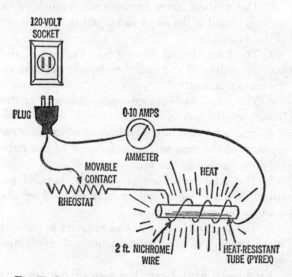

120-VOLT
SOCKET

PLUG

0-10 AMPS

AMMETER

MOVABLE
CONTACT

HEAT

RHEOSTAT

2 ft. NICHROME
WIRE

HEAT-RESISTANT
TUBE (PYREX)

Fig. 70. Setup for Experiment 15: Heat from Electricity

Now turn the sliding contact of the rheostat to the extreme *left* (for maximum resistance) and plug the cord into the power outlet. As you can confirm with an ammeter inserted into the line, only a small current flows and, hence, little heat is produced and radiated from the Nichrome heater wire. Move the

slider on the rheostat progressively more to the *right* to reduce the series resistance and increase the current through the circuit. Observe that the heat produced by the nichrome wire becomes more and more *intense* and that the wire first becomes dull *red* in color, then *orange* and finally *yellow* in color, indicating rapidly increasing temperatures. Do *not* continue the experiment beyond yellow heat of the nichrome wire; greater currents may overload the rheostat and will blow a fuse before burning out the resistance wire.

Heat and Work: Joule's Law. The heat produced by the resistance wire in the experiment just concluded is a measure of the work done by the electric current in overcoming the resistance of the conductor. The energy required for this work is supplied by a battery, generator, or other current source. The more heat is produced, the more work is performed by the current and, hence, the more energy is expended by the battery or generator. If we could find out, then, *how much* heat is generated by an electric current, we would also know how much energy had been expended and vice versa. (This is true *provided* the electric current produces *only* heat and performs *no other* chemical or mechanical work.)

The English physicist JAMES PRESCOTT JOULE (1818-1889) concerned himself with this problem and in 1840 published a famous paper "On the Production of Heat by Voltaic Electricity," which contained the results of his experiments. Based on these experiments, he announced a law (Joule's Law), which stated that **the total amount of heat developed in a conductor is directly proportional to the resistance, to the square of the electric current, and to the length of time during which the current flows** (i.e., the wire is heated). Expressed as an equation, Joule's law of electric heating gives the amount of heat (H) produced in a time, t, by a current, I, in a conductor of resistance, R, as

$$H = I^2 R t \text{ (joules)} \qquad (1)$$

where H is the heat energy in **joules**, if the current is in amperes, the resistance in ohms, and the time in seconds.

While Joule's law was arrived at experimentally and has been confirmed by innumerable subsequent experiments, it is easy to show its theoretical validity by fundamental energy considerations. Earlier we defined the potential difference (V) between two points as the work (W) done in carrying a unit charge through a conductor from the point of high potential to the point of lower potential. We also found that the total work (W) done in moving any charge (Q) between the points is simply the *product* of the charge and the potential difference, or

$$W = Q \times V \qquad (2)$$

Moreover, since the charge (Q) is the product of the current (I) and the time (t), we may write

$$W = Q \times V = I \times t \times V \qquad (3)$$

Finally, by Ohm's law, the potential difference V = I R, and hence

$$W = I \times t \times I R = I^2 R t \qquad (4)$$

Equation (4) expresses the total amount of work done or the energy expended *by* an electric current. By the principle of conservation of energy, the **electrical energy (W) expended must equal the heat energy (H) produced,** or

$$W = H = I^2 R t \qquad (5)$$

which is identical with Joule's law (eq. 1).

Joule's law (eq. 5) gives the *energy in joules.* Since *heat* is usually measured in *calories,* we would like to know how many calories of heat are produced for each joule of energy. Joule determined this **electrical equivalent of heat** in 1841, stating that the heat in calories = 0.239 × energy in joules; and equivalently, energy in joules = 4.18 × heat in calories. (You obtain the second equation by dividing through by 0.239; the reciprocal of 0.239 is 4.18). Applying this result to equation (1) or (5), we obtain finally for the heat (H) produced by a current:

$$H \text{ (calories)} = 0.239 \, I^2 R t = 0.239 \, V I t \qquad (6)$$

where *I* is the current in amps, *R* is the resistance in ohms, *V* is the potential difference (voltage), and *t* is the time in seconds.

EXAMPLE: An electric heater, operating from the 120-volt power line, consists of two coils of resistance wire, each with a resistance of 30 ohms. The coils may be connected either in series or in parallel. Compute the heat generated in 10 minutes for either case.

Solution: When the coils are connected *in series,* their total resistance is 60 ohms. The current, therefore, is $\dfrac{E}{I} = \dfrac{120}{60} = 2$ amps. Substituting in eq. (6) above, the heat in calories

H = 0.239 I² R t = 0.239 (2 × 2) × 60 × (10 × 60) = 34,400 cals.
Alternatively, H = 0.239 V I t = 0.239 × 120 × 2 × 600 = 34,400 calories.

When the coils are connected in parallel, their equivalent resistance R = 30/2 = 15 ohms. The current I = E/R = 120/15 = 8 amps. Hence, the heat produced (in calories)

H = 0.239 I² R t = 0.239 (8)² × 15 × 600 = 137,900 calories.
or H = 0.239 V I t = 0.239 × 120 × 8 × 600 = 137,900 calories.

Thus, the parallel connection produces *four times* as much heat as the series connection.

Joule's Law for A.C. The heating effect of an electric current does *not* depend on the type of current. Joule's law is equally valid for *direct and alternating* current. As we shall see later, an alternating current (a.c.) continuously reverses its polarity from plus (+) to minus (−) and back to plus. But the *square* of the current, which enters into Joule's law, remains the same regardless of the polarity; thus $(+I)^2 = (-I)^2 = I^2$. The fact that the heating effect is the same for all types of current is made use of in some electrical measuring instruments that measure both a.c. and d.c. It is also the basis for comparing the *effective* values of alternating and direct currents.

ELECTRIC POWER: THE RATE OF DOING WORK

Power, whether electrical or mechanical, is always the **rate of doing work.** Alternatively, power is the **work accomplished per unit time.** We have already computed the work (W) performed by an electric current in flowing from a point of high potential to a point of lower potential through a conductor of resistance R. By equation (4)

$$W = I^2 R t$$

The power (P) is the *time-rate* of doing work, or

$$P = \frac{W}{t} = I^2 R \qquad (7)$$

Where *P* is expressed in joules/second, a unit which is familiarly known as the **watt.** Large amounts of power are usually expressed in **kilowatts** (abbreviated *kw*). One kilowatt equals 1000 watts.

By Ohm's law, we may substitute for the current I = E/R in (7).

Hence, $P = I^2 R = \left(\dfrac{E}{R}\right) \times I \times R = E \times I$ (8)

or $P = I^2 R = (E/R)^2 \times R = \dfrac{E^2}{R}$ (9)

All three forms of the power formula $\left(P = I^2R = EI = \dfrac{E^2}{R}\right)$ are constantly used and apply to all circuits governed by Ohm's law.

EXAMPLE 1: What is the power required to drive a motor that has a current consumption of 15 amperes at an applied voltage of 120 volts?

Solution: Power $= E \times I = 120 \times 15 = 1,800$ watts $= 1.8$ kilowatts

EXAMPLE 2: A generator with a terminal voltage of 220 volts sends a current of 0.5 ampere through a lamp having a resistance of 440 ohms. Compute the power required by the lamp.

Solution: The power may be computed in *three* different ways:

(1) Power in lamp $= E \times I = 220 \times 0.5 = 110$ watts

(2) Power in lamp $= I^2 R = (0.5)^2 \times 440 = 0.25 \times 440 = 110$ watts

and (3) Power in lamp $= \dfrac{E^2}{R} = \dfrac{(220)^2}{440} = \dfrac{48,400}{440} = 110$ watts.

How to Figure Electric Power Costs. When you pay your electric bill, you are being charged for the *work performed, not for the rate* of doing it. In other words, you are paying for *energy* consumed, *not for power.* By eq. (7), the work or energy equals the **product of power and time,** or

Energy $= P t = E I t = I^2 R t = \dfrac{E^2}{R} t$ (10)

where the power is in watts, the voltage in volts, the current in amps, and the time is in seconds. The energy thus comes out in joules, or *watt-seconds.*

Electric power companies charge for energy in units of **kilowatt-hours,** rather than watt-seconds. To obtain the energy in kilowatt-hours, you must divide the power in watts by 1000 (move the decimal point three places to the left) and then multiply the result by the number of *hours* energy is being consumed.

EXAMPLE: Five 100-watt bulbs, a 300-watt radio, and a 1200-watt air conditioner are operating from 8 A.M. to 6 P.M. How much energy in kilowatt-hours is consumed?

Solution: The total power $P = 5 \times 100 + 300 + 1200 = 2000$ watts $= 2$ kw. Hence, the energy $= P t = 2 \times 10$ hrs $= 20$ kw-hours.

The term "horse-power" is frequently used to state the power of large motors, air conditioners, etc. To convert horse-power (abbreviated H.P.) to watts or kilowatts, memorize the relation

1 horse-power $= 746$ watts $= 0.746$ kilowatt

EXAMPLE: A ¾-H.P. air conditioner operates from the 115-volt power line for 24 hours. How much energy does it consume? What current does it draw?

Solution: ¾ H.P. $= ¾ (746) = 560$ watts $= 0.56$ kw. Hence, the energy $= P t = 0.56 \times 24 = 13.44$ kilowatt-hours.

The current $I = \dfrac{P}{E} = \dfrac{560}{115} = 4.87$ amperes.

Most electric appliances have the **wattage rating** either directly stamped upon them, or they indicate the **current** drawn by the device. If the current is indicated, you must multiply by the **line voltage** to obtain the wattage or power rating. Electric bulbs and household appliances are always marked with the wattage rating. Electric motors, in contrast, usually have the number of amps drawn and sometimes the horse-power rating marked on the nameplate. Table VI lists the range of wattage and ampere ratings for some typical household appliances. The wattage rating will help you to figure your electric bill, while the current rating (added up for all the appliances connected to a common line) will aid you in selecting the right gauge wire.

TABLE VI

POWER CONSUMPTION AND AMPERE RATINGS FOR HOUSEHOLD APPLIANCES

Electric Appliance	Range of Power Consumption (in watts)	Current Range for 115-Volt Line (in amperes)
Home Air Conditioners	850–1700	7.5 –15
Electric Blankets	100–500	0.87 – 4.3
Coffee Makers	660–1000	5.7 – 8.7
Deep Freezer	150–400	1.3 – 3.5
Frying Pan	1300	11.3
Grill	500–1200	4.3 –10.45
Electric Heaters	450–1500	3.9 –13
Heating Pads	40–60	0.4 – 0.5
Hot Plates	550–1400	4.8 –12
Flat Irons	550–1200	4.8 –10.45
Projectors (Movie,		

Slide)	300–1000	2.6 – 8.7
Pressure Cookers	500–1200	4.3 –10.45
Radios (Receivers, Hi-Fi)	50–500	0.435– 4.35
Refrigerators	100–300	0.87 – 2.75
Rotary Broilers	1000–1400	8.7 –12
Sewing Machine	30–60	0.26 – 0.53
Electric Shavers	10–25	0.087– 0.22
Soldering Irons	30–750	0.26 – 6.5
Television Sets	150–550	1.3 – 4.8
Toasters	750–1300	1.6 –11.2
Waffle Irons	600–1000	5.2 – 8.7
Automatic Washers	200–700	1.7 – 6.1
Dryers	800–3000	7–26

To calculate your electric bill you must *multiply the total number of kilowatt-hours* of energy consumed (for a stated period) *by the rate*, in cents per kilowatt-hour (kw-hr). This rate is usually staggered, less being charged the more kilowatt-hours you consume. Let us see how this works out for the example of a typical monthly bill.

Sample Calculation: The monthly rate chart of one power company looks like this:

Energy Consumed	Rate
First 20 kilowatt-hours (or less) $1.60	
Next 100 kw-hrs	4.5 cents per kw-hr
Next 100 kw-hrs	3.5 cents per kw-hr
Next 100 kw-hrs	3.0 cents per kw-hr
All over 320 kw-hrs	2.0 cents per kw-hr

Let us assume that in this household eight 75-watt bulbs burn on an average of 5 hours a day; a 300-watt radio operates 3 hours a day; a 500-watt television set is on for 4 hours a day; a ¾ H.P. air conditioner operates 8 hours a day; an 800-watt toaster for ½ hour each day; a 1000-watt broiler 1 hour a day, and miscellaneous sources (shavers, clocks, etc.) totaling up to 100 watts operate 6 hours a day. What is the total power consumption and the bill?

We calculate as follows: (See below).

HEAT IN ELECTRICAL CIRCUITS AND APPARATUS

Heating occurs in all electrical circuits and is proportional to the product of the *square* of the current and the resistance (I^2R). In most electrical apparatus this heating effect is undesirable, though unavoidable. We cannot prevent the windings of motors and transformers, from heating up and thus wasting part of the energy supplied. We must make sure, however, that most of the heat is carried away, either naturally or by forced cooling, to prevent the device from overheating and possibly being damaged. The power (wattage) rating of an electrical device, in addition to informing us of the cost of its operation, is also an important indicator of its safe operation. The power rating indicates the maximum permissible heating of the device, and it must not be exceeded if damage is to be avoided.

In some electrical devices, the heating effect is deliberately used for a specific purpose. Thus fuses, consisting of a piece of high-resistance wire of low melting point, are designed to "blow out" when the current in a circuit exceeds the maximum permis-

Appliance	Power (kw)	Energy per Day (kw-hrs)	Energy per Month (kw-hrs)
eight 75-w bulbs	8 × 75 = 600 = 0.6 kw	0.6 × 5 = 3 kw-hrs	3 × 30 = 90
300-w radio	300 watts = 0.3 kw	0.3 × 3 = 0.9	0.9 × 30 = 27
TV set	500 w = 0.5 kw	0.5 × 4 = 2	2 × 30 = 60
¾-H.P. Air Cond.	¾ × .746 = 0.56 kw	0.56 × 8 = 4.48	4.48 × 30 = 134
toaster	800 w = 0.8 kw	0.8 × ½ = 0.4	0.4 × 30 = 12
broiler	1000 w = 1 kw	1 × 1 = 1 kw-hr	1 × 30 = 30
Miscellaneous	100 w = 0.1 kw	0.1 × 6 = 0.6	0.6 × 30 = 18
Totals:	3.86 kw	12.38 kw-hrs/day	371 kw-hrs/month

The total power consumption, thus, is *371 kw-hrs per month.*

Hence, the first 20 kw-hrs .... $ 1.60
The next 100 kw-hrs @ 4.5¢ per kw-hr 4.50
The next 100 kw-hrs @ 3.5¢ per kw-hr 3.50
The next 100 kw-hrs @ 3.0¢ per kw-hr 3.00
And 51 kw-hrs @ 2.0¢ per kw-hr = 51 × .02 .... 1.02

Total   371 kw-hrs   Total $13.62

sible value and *before* other equipment can be over-loaded or damaged. Electric *heaters* rely exclusively on the I²R effect and have no other purpose but to convert electricity into the maximum amount of heat. The heating of the wire filament in *radio tubes* makes possible the **emission of electrons**, on which the operation of the tubes is based. In *incandescent* and *arc lamps* heating always accompanies the production of light, though engineers strive to make the ratio of light to heat as large as is feasible. (In some types of "cold" and fluorescent lamps, light is produced with practically no accompanying heat.)

Fuses. The purpose of a fuse is to protect electrical equipment, motors, instruments, radio receivers, etc., from *excessive currents* resulting from sudden "overloads" or accidental "shorts." A fuse is always connected *in series* with the apparatus it is to protect, so that it will open *before* the rest of the circuit is injured. Remember that a fuse cannot *correct* a faulty circuit condition, but merely *opens* the circuit, thus preventing further damage. It does this by melting or "blowing out." You can tell when a fuse is blown by the broken wire strip (filament) and the darkened glass. A fuse must blow very quickly in the event of a large overload or a **short** circuit. (A number 18 household wire, for example, will vaporize in 1/100 second by a short-circuit current of 2000 amperes.) Yet fuses must operate fairly slowly in the event of moderate or **momentary overloads**. Any fair-sized electric motor, for example, draws a **starting current** considerably higher than its normal operating current. The fuse must operate slowly enough *not* to interrupt the motor circuit for the momentary overload during the

starting period. Special dual-element **time dela** fuses have been designed for this purpose.

Fig. 71 illustrates various types of fuses in com mon use. Most of these contain zinc-strip element of relatively high resistance and low melting poin The screw-plug fuse is familiar in every household The cartridge types, designed from low to ver high current capacities, must be inserted into a appropriate **fuse holder**. The **renewable link** fuse may be inserted directly between two terminals and are easily replaced. As shown in the illustration the renewable link fuse melts at two narrow con strictions when the permissible current is exceeded

Incandescent Lamps. Incandescent lamps consis of a thin, resistive filament wire in series with tw lead-in wires that are soldered to a screw-type o bayonet (plug-in) base. The incandescent filamen is mounted in a highly evacuated glass bulb to which some argon and nitrogen gas is added. The presence of the gases permits higher operating temperatures without danger of the filament being melted. The early lamps (the type invented by Edison) used relatively inefficient carbon filaments, bu all present-day lamps use drawn **tungsten wire** filaments. In some bulbs of low wattage the filament may be a fraction of a human hair in width. The operating temperatures of incandescent lamp filaments extend over a range from about 3600°F to 5900°F, the upper limit being reached by the very bright, short-lived photoflood lamps.

(By the way, the wattage rating of an incandescent lamp applies only to **parallel connection** of lamps across the power line. If you connect lamps *in series*, the wattage rating may completely fool you. For example, a 10-watt bulb has a thin hair-like filament, while a 1000-watt bulb, in contrast has a relatively heavy filament of much lower resistance. If you connect a 10-watt and a 1000-watt bulb *in series* across the 115-volt line, the current will heat up the high-resistance filament of the 10-watt bulb (heat = I²R), while hardly affecting that of the 1000-watt bulb. Thus, the "power" of the 10-watt bulb is much greater than that of the 1000-watt bulb in this case.)

Electric Arc Lamps. If the tips of two carbon rods, through which a current of about 15 to 20 amperes is passing, are first brought together and then slightly separated, the current will continue to flow as a high-intensity **arc discharge**. Moreover, the tips of the rods will become extremely hot, reaching temperatures of 4000 to 6000°C, and an intense light will be projected, as well as ultra-

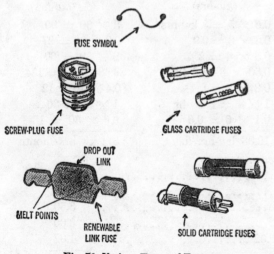

FUSE SYMBOL

SCREW-PLUG FUSE

GLASS CARTRIDGE FUSES

DROP OUT LINK

MELT POINTS

RENEWABLE LINK FUSE

SOLID CARTRIDGE FUSES

Fig. 71. Various Types of Fuses

violet and infrared radiation. What happens is that the high-resistance carbon tips vaporize at the contact point and the vaporized particles form a **conducting path** for the arc after the rods are separated. The length of the arc must be adjusted continuously, since the arc extinguishes if the separation between the tips becomes too great. Adjustment is usually done automatically.

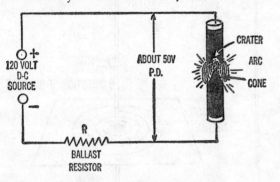

Fig. 72. Circuit of Electric Arc Light

Fig. 72 illustrates the schematic circuit diagram of an arc light operating on 120 volts D.C. A series ballast resistor is required to reduce the line voltage to about 50 volts across the arc and to limit the maximum current. With D.C., the *positive* rod becomes very hot by electron bombardment from the negative rod (cathode) and forms a small *crater*. The negative rod, in contrast becomes *cone-shaped*. Electric arcs do *not* obey Ohm's law. The potential difference across the arc *drops* as the current *increases*.

EXAMPLE: The arc of Fig. 72 requires a voltage of 50 volts and a current of 10 amps. What is the value of the ballast resistor? The wattage?

Solution: $R = \dfrac{E}{I} = \dfrac{120 - 50}{10} = \dfrac{70}{10} = 7 \ ohms.$

The power rating
$P = I^2 R = (10)^2 \times 7 = 100 \times 7 = 700 \ watts.$

Practice Exercise No. 8

1. Pieces of copper and iron wire of equal length and equal thickness are connected first in series and then in parallel with a low voltage source. Which wire develops the greater heat in each case?

2. Find the total heat produced by an electric iron drawing 5 amperes at 120 volts for a period of 20 minutes.

3. An electric toaster takes a current of 7 amps at 120 volts. How much heat is given off per hour? What is the cost per hour at the rate of 5¢ per kilowatt-hour?

4. A neon electric sign consumes 4.8 watts per foot of tubing and has a total length of 135 feet. How much does it cost to operate the sign per week, if it is on 10 hours a day, and electricity costs 5¢ per kw-hr?

5. What is the power consumption of the circuit described in Problem #8 of Practice Exercise No. 6?

6. A factory is charged for electrical *energy* and for *maximum power demand* according to the following *monthly* schedule:

Energy	Power
First 1000 kw-hr: 5¢/kw-hr	First 50 kw..$2.40 per kw
Next 4000 kw-hr: 3¢/kw-hr	Next 750 kw.$2.00 per kw
Next 50,000 kw-hr: 1.3¢/kw-hr	All above this.$1.50 per kw
All above this total: 1¢/kw-hr	

What will be the electric bill of the factory for 75,000 kw-hrs of energy during a month, when the maximum power demand was 300 kw?

7. Compute the resistance value and wattage of the ballast in the circuit of Fig. 72, if the current is 15 amps and the arc voltage is 40 volts, for a line voltage of 115 volts.

SUMMARY

Heat is produced whenever an electric current flows through a resistance. **Joule's Law:** The total amount of heat developed in a conductor is directly **proportional to the resistance**, the **square of the current**, and to the **duration of the current** ($H = I^2 R t$ joules).

The **heat in calories** = $0.239 \times$ **energy in joules**
or H (calories) = $0.239 \ I^2 \ R \ t = 0.239 \ V \ I \ t$

Joule's law holds for **direct** as well as **alternating** current.

Power is the rate of doing work, or the work accomplished per **unit time.** $P = I^2 R = E I = \dfrac{E^2}{R}$ (watts).

Energy is the **product of power and time** ($P \times t$) and is usually expressed in **kilowatt-hours** (1 kilowatt = 1000 watts).

One **horse-power** equals 746 watts (or 0.746 kilowatt).

To figure electric energy costs, multiply the total number of kilowatt-hours by the *rate* (in cents) per kilowatt-hour.

Fuses are connected **in series** with electrical equipment to protect it from excessive currents caused by overloads and shorts.

Electric arcs are formed by the vaporization of carbon particles at the heated carbon tips. Arcs do *not* obey Ohm's Law. A **ballast resistor** must be connected **in series** with the arc to limit the current and obtain the required voltage drop across the arc.

CHAPTER NINE

ELECTROMAGNETISM—CHARGES IN MOTION

OERSTED'S DISCOVERY: MAGNETIC EFFECT OF ELECTRIC CURRENT

The Danish physicist HANS CHRISTIAN OERSTED discovered during a lecture demonstration in 1820 that a compass placed near a current-carrying wire was deflected from its normal North-South direction to a position nearly perpendicular to the wire (see Fig. 73). When the direction of the current was reversed, the needle would deflect in the opposite direction, again assuming a position almost at right angles to the wire. Oersted noted that the deflection of the needle lasted only as long as current was flowing through the wire and hence could not be caused by the (copper) wire, but must be due to the current itself. The fundamental and far-reaching fact that an **electric current** (i.e., charges in motion) is always surrounded by a magnetic field was thus discovered quite accidentally.

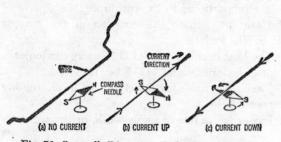

Fig. 73. Oersted's Discovery: Deflection of Magnetic Compass Needle by an Electric Current

In investigating further, Oersted found that the *direction* of the magnetic field is *perpendicular* to the wire and that its intensity diminishes as the distance from the wire is increased. Thus, if a wire is connected to a battery and placed in a vertical position, a nearby compass needle will always tend to set itself at right angles both to the wire and to the perpendicular from the wire to the center of the needle. (See Fig. 74.) If the needle is carried in a circle around the wire, as indicated in Fig. 74, the axis of the needle will always be *tangent* to the circle. Moreover, if the (electron) current is *up* through the wire, as shown, the N-pole of the needle will always point in a clockwise direction, as the needle is carried around the wire.

If the compass is moved around the wire at a closer distance, the needle (specifically, its axis)

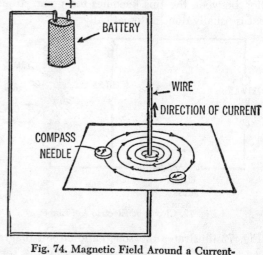

Fig. 74. Magnetic Field Around a Current-Carrying Wire

will trace out another circle of smaller radius; if the compass is moved around the wire at a larger distance, the needle will trace out a circle of greater radius than before. The magnetic field lines around a wire thus consist of a series of **concentric circles** (Fig. 74), indicating the way a north magnetic pole would move about the wire, if free to do so. Fig. 75 illustrates the lines-of-force representations of the magnetic field around a current-carrying wire. In (*a*) of the figure, the electron current flows from *left to right into the page* (away from you), as indicated by the cross in the plane representation. The field lines in this case are *counterclockwise*, as shown by the arrows. In (*b*) the electron current is flowing *out of the page* (towards you), as indicated by the dot in the plane representation. The direction of the field lines in this case is *clockwise*, indi-

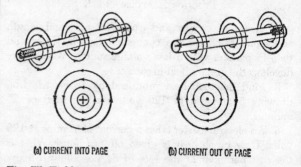

(a) CURRENT INTO PAGE   (b) CURRENT OUT OF PAGE

Fig. 75. Field Representations Showing Relative Directions of Electron Current and Lines of Force

cating that a *free* north pole would move in that direction.

Left-Hand Rule for Conductors. Fig. 76 illustrates the simple rule for determining the relative directions of the (electron) current and the lines of force: *Grasp the current-carrying wire with the left hand, with the thumb pointing in the direction of the electron flow along the wire; when the fingers are wrapped around the conductor, they will point along the lines of force.* Remember that this rule applies to **electron flow**; for *conventional* current, which flows in the opposite direction, the *right* hand is used to determine the direction of the field lines.

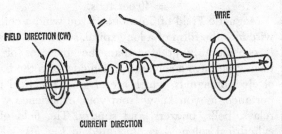

Fig. 76. Left-Hand Rule for Determining the Direction of the Field Lines Around an Electron Current Carrying Conductor

EXPERIMENT 16: You can demonstrate to your own satisfaction that a magnetic field exists around a current-carrying conductor by the following simple experiment. Make a test stand with heavy copper wire or tubing, as shown in Fig. 77. Connect the ends of the stand with wires to a battery made up of three dry cells in series. If a switch is available, connect it in series with one of the connecting wires, or leave the wire disconnected from one of the battery terminals until actually starting the experiment. Place a cardboard (or Lucite sheet, if available) through a central hole over the test stand and hold it in a horizontal position. Sprinkle some iron filings evenly over the surface of the cardboard or Lucite sheet.

Now close the switch or connect the end of the wire to the free battery terminal. Tap the cardboard lightly with your finger and observe how the iron filings arrange themselves in concentric circles along the lines of force. (Caution: Do not leave the current on for any length of time to avoid exhausting the battery, which is shorted.) If you now open the switch and again tap the cardboard, the iron filings will resume their random distribution on the board. To show that the circular pattern is actually the result of the magnetic field about the conductor, repeat the experiment by closing the switch and

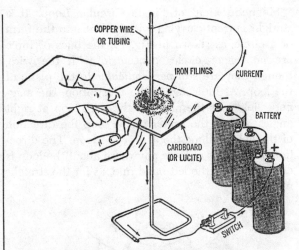

Fig. 77. Test Setup for Experiment 16.

tapping the cardboard. The iron filings will immediately arrange themselves in the field pattern of Fig. 74.

Magnetic Field of a Straight Conductor. A short time after Oersted's discovery, two French physicists, JEAN B. BIOT (1774-1862) and FELIX SAVART (1791-1841), made some measurements on the intensity of the magnetic field at various distances from a long current-carrying wire. They found that **the field intensity varied directly with the strength of the current through the conductor and inversely with the distance from the point of measurement to the nearest point on the wire.** Expressed quantitatively, the **Law of Biot and Savart** states that the field intensity H (in oersteds) at a point r cms from a long straight wire carrying a current of I amps, is

$$H = \frac{2I}{10\,r} \text{ oersteds}$$

EXAMPLE: What is the field intensity at a point 5 cms from a long wire carrying a current of 20 amperes?

Solution: $H = \dfrac{2I}{10\,r} = \dfrac{2 \times 20}{10 \times 5} = 0.8$ oersted.

MAGNETIC FIELD ABOUT CIRCULAR LOOPS: SOLENOIDS AND ELECTROMAGNETS

The field about a straight wire is weak. Much stronger fields are obtained by coiling wire into a spiraling loop, known as a **solenoid.** Before going into solenoids and the electromagnets evolved from them, let us see what happens to the field about straight wire, when the wire is bent into the form of a circular loop.

Magnetic Field of Single Circular Loop. If a straight current-carrying wire is bent into the form of a circle, as shown in Fig. 78, the lines of force are no longer circles concentric with the wire, though they are still *perpendicular* to the plane of the loop. As indicated in the illustration, the magnetic field cuts the plane of the wire loop at right angles, and at the *center* of the loop the direction of the field is along the axis of the loop. The direction of the lines of force anywhere about the loop is determined by the left-hand rule, as for the straight wire.

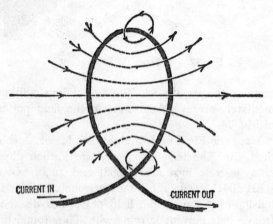

CURRENT IN   CURRENT OUT

Fig. 78. Magnetic Field About a Circular Loop

It is sometimes of interest to know the magnetic field intensity at the *center* of such a current-carrying loop. By use of the calculus it may be shown that the field intensity H at the *center* of such a loop of radius *r* (cms) and carrying a current of *I* amperes, is

$$H = \frac{2 \pi I}{10 r} \text{ oersteds}$$

where $\pi = 3.14159$

By comparing this equation with the one given before for a straight conductor, you can see that the simple expedient of coiling the wire into a loop has increased the magnetic field intensity (at the center) by a factor of π or 3.14.

The field intensity equation for a circular loop is sometimes used to **define the ampere** as that current which in a circular loop of 1 cm radius will produce a field intensity of $2\pi/10$ or 0.62831 oersted at the center.

Magnetic Field of Flat Coil. If you wind several loops of wire close together into a **flat coil** and pass a current through all the loops, the magnetic field about each turn will have the same direction and each loop will contribute an equal amount to the total field intensity at the center. If the thickness (length) of the coil is *small compared to its radius*, a flat coil of *N* turns will have a field intensity at its center *N* times as great as that of a single loop, or for a flat coil

$$H = \frac{2 \pi N I}{10 r} \text{ oersteds (at center)}$$

EXAMPLE: What is the magnetic field intensity at the center of a flat coil of 40 turns, if the radius of the coil is 6.28 cms and it carries a current of 10 amperes?

$$\textit{Solution: } H = \frac{2 \pi N I}{10 r} = \frac{2 \times 3.14 \times 40 \times 10}{10 \times 6.28}$$
$$= 40 \text{ oersteds.}$$

Magnetic Field of Solenoid. If you wind a coil of wire into the form of a long spiral, so that its length is considerably greater than the diameter of its turns, you have a solenoid. Solenoids form the basis of **electromagnets**, which are of great practical importance, as you know from your experience with relays, bells, buzzers, and sundry. The field of a cylindrical solenoid is illustrated in Fig. 79.

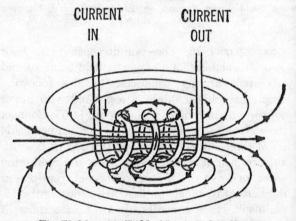

CURRENT IN   CURRENT OUT

Fig. 79. Magnetic Field of Long Coil (Solenoid)

With current flowing through the coil, the lines of force leave the solenoid at one end, the north pole, and enter it at the opposite end, the south pole. You can determine the direction of the magnetic field in a solenoid with the **modified left-hand rule,** illustrated in Fig. 80: *Wrap the fingers of the left hand around the coil in the direction of the electron current flow; the thumb will then point in the direction of the north pole of the coil,* from which the lines of force leave.

Note in Fig. 79 that while all the lines of force pass through the *center* of the coil, some of the lines do *not* return through the outside path to the *ends* of the coil, but cut through the coil at inter-

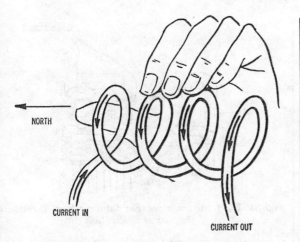

Fig. 80. **Left-Hand Rule for Long Coils (Solenoids)**

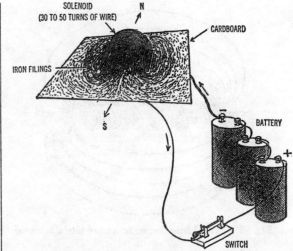

Fig. 81. **Experiment 17: Magnetic Field About a Solenoid**

mediate points. Because of this magnetic *leakage* the magnetic field intensity near the ends of the coil is not as great as near the central portion. The formula for the field intensity of a solenoid *near its center*, which is of greatest interest, is as follows:

$$H = \frac{4 \pi N I}{10 \, l} \text{ oersteds}$$

where l is the length of the coil in cms, N is the number of turns, and I is the current in amperes. The formula is valid only for a coil whose length is great in comparison to its cross section.

EXAMPLE: What is the magnetic field intensity near the center of a solenoid of 100 cms (39.37 in.) length and having 2000 turns, when a current of 5 amperes flows through the coil?

Solution: $H = \dfrac{4 \pi N I}{10 \, l} = \dfrac{4 \times 3.1416 \times 2000 \times 5}{10 \times 100}$.

$$= 125.9 \text{ oersteds.}$$

Let us perform a simple experiment to illustrate further the field about a solenoid.

EXPERIMENT 17: Cut out a cardboard so that you can fit a solenoid in it, with half projecting above and half projecting below the cardboard (see Fig. 81). Wind the coil with 30 to 50 turns of bell wire. Connect the ends of the coil through a series-connected switch (if available) to a battery, similar to the setup for Experiment 16.

If you now close the switch (or connect the free end of the wire to the battery) and sprinkle some iron filings on top of the cardboard, you will see the filings arrange themselves in the pattern illustrated in Fig. 79. (You will have to tap the cardboard lightly to aid the formation of the pattern.) Note that this field pattern is essentially the same as that of an ordinary bar magnet, illustrated in Fig. 23.

You can extend the experiment if you have a tiny compass needle available. Remove the iron filings and place the compass needle inside the solenoid near its center. The needle will align itself with the axis of the coil and its N-pole will point toward the north pole of the coil, as given by the left-hand rule.

Magnetic Field of Ring Solenoid. A ring solenoid is a long coil bent into the form of a ring or **toroid,** and looks as if wire were wound around an automobile tire. The magnetic field inside a ring solenoid is confined entirely inside the loops of wire and does not project outside the ring. The formula for the long, cylindrical coil or solenoid also gives the magnetic field intensity *inside* the loops of a ring solenoid.

Electromagnets and Their Applications. The field of an **air-core** solenoid, the type we have studied, is still relatively weak. By inserting a **core of soft iron** into a solenoid the number of lines of force per unit area and, hence, the **flux density,** can be greatly **increased for the same magnetizing force** applied to the solenoid (see Fig. 82). This increase in flux density is accomplished by the large number of *additional* lines of force produced by the temporary magnetization of the iron. You will recall that the relative ease of magnetization of a material compared to air (specifically, to a vacuum) is given by the **permeability** of the material. The permeability of soft wrought iron is *several thousand times that of air,* which explains the large increase in magnetic flux, when an iron core is inserted into a solenoid.

A coil of wire wound around a soft-iron core is called an **electromagnet.** Electromagnets in various forms surround us in thousands of diverse applica-

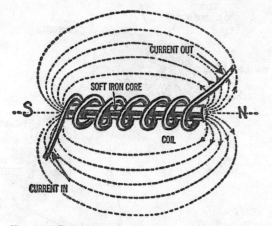

Fig. 82. Effect of Inserting an Iron Core into a Solenoid

tions. Electromagnets energize the fields of motors and generators. Powerful "lifting magnets" hold tons of scrap iron and machine parts by simply closing a switch and drop the load, when the switch is opened. Electromagnets are part of telephones, loudspeakers, buzzers, electric bells, telegraphs, relays, electric meters, and many other devices.

If you bend an iron core into the form of a horseshoe and wind a coil of wire on each leg, you can obtain a powerful electromagnet. Connect the two coils of wire in series, the end of one to the beginning of the other, as illustrated in Fig. 83, and apply

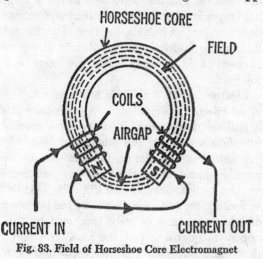

Fig. 83. Field of Horseshoe Core Electromagnet

a d-c voltage to the free ends. As you can check with the left-hand rule for coils, the fields of the two coils are in the same direction and thus *aid each other.* (This is known as series-aiding.) As a consequence, a concentrated magnetic flux travels around the horseshoe and across the airgap between the two poles (or legs). The shorter you make this airgap, the more concentrated is the flux density between the poles. Such a horseshoe electromagnet

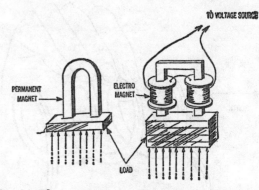

Fig. 84. Electromagnets are Far Stronger than Permanent Magnets of Similar Size

has a far greater lifting ability than a permanent magnet of equal size, as illustrated by the different iron loads applied to each magnet in Fig. 84.

The Relay. An electromagnetic relay permits a weak current in one circuit to control a heavy current flowing in another circuit. A relay is thus essentially a switch that permits closing a circuit at some remote location. Relays are used in thousands of control applications, where switches are not practical.

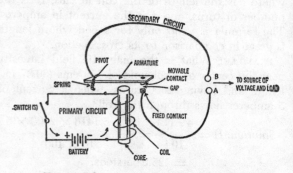

Fig. 85. Schematic Presentation of Relay and Associated Circuits

As illustrated in schematic form in Fig. 85, a relay contains an iron-core electromagnet, which is mounted close to a flat blade of magnetic material, called **armature.** The armature is pivoted at one end and is held a small distance away from the core of the electromagnet by means of a spring. A contact made of a good conductor, such as silver or tungsten, is attached to the free, movable end of the armature. A second, stationary contact is mounted opposite the movable contact, separated by a small distance, known as **contact gap.**

When the switch S is closed in the primary circuit, current flows from the battery through the coil of the electromagnet and magnetizes the core. As a consequence the armature is attracted to the core of the magnet and thus closes the contacts of the **secondary circuit.** Current then flows through this secondary circuit, consisting of a voltage source and load attached between terminals *A* and *B*. If the switch (*S*) in the primary circuit is opened, the spring returns the armature to its original position, thus breaking the contacts of the secondary circuit. The contacts in the relay shown are *normally open,* when the relay is not energized. Contacts may also be arranged to be *normally closed,* when the relay is de-energized.

The Telegraph. Relays are used as part of the conventional **telegraph** system. A typical telegraph station is illustrated in Fig. 86. When a telegraph

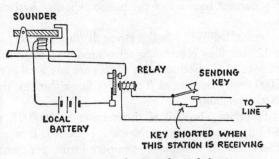

Fig. 86. Essentials of a Telegraph Station

sending key at some remote station is closed, a pulse of current flows through the long-distance telegraph line and through the shorted key of the local station to energize a local relay. The contacts of the relay then close a secondary circuit at the local station and current flows from the local battery through the relay contacts to actuate a **sounder.** The sounder is another electromagnetic device that produces a distinctive click each time the heavy bar armature is attracted by an electromagnet. A succession of such signals spells out the telegraphic message. If the local station wants to send a message, the short across its sending key is removed and the same process takes place at the remote station, as was just described for the local station.

In the **teletype system,** used by police, news bureaus and business firms, the key and sounder are replaced by machines similar to an electric typewriter. As the message is typed out at one station, it is reproduced on paper tape by the other **teletype machine** at the remote end of the line.

The Electric Bell. The electric door bell or buzzer is the most familiar device using an electromagnet. The buzzer or bell resembles a relay, but it uses *interrupted* magnetization to produce an audible signal tone. As is illustrated in Fig. 87, the heart with many turns of fine wire wound around each of a door bell is a horseshoe-type electromagnet,

leg of the magnet. In contrast to the relay, the armature is connected directly to one end of the coil and its normal resting position is against an adjustable, external contact point.

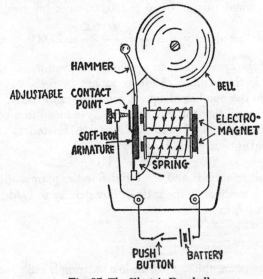

Fig. 87. The Electric Doorbell

When the pushbutton is pressed, the circuit is closed and current flows from the battery through the electromagnet, the armature and the contact point, back to the battery. As a consequence, the core of the magnet is magnetized and attracts the soft-iron armature, thus striking the bell with the attached hammer. As soon as the armature moves, the contact opens and the circuit is interrupted. The magnet then becomes de-energized and the spring pulls the armature back against the contact, thus closing the circuit once more. This action then repeats at a rapid rate, making the hammer strike the bell many times each second. In a buzzer the hammer and bell are left off, so that the vibration of the armature alone produces the characteristic buzzing sound. A set screw permits adjusting the contact gap between the contact point and the armature, which in turn adjusts the buzzer pitch or bell tone.

MAGNETIC CIRCUITS AND HYSTERESIS

Let us now consider some of the factors that control the magnetic flux (strength) of electromagnets and other **magnetic circuits.** First let us recall some definitions. The total number of lines of force in a magnetic field is called the **magnetic flux** (symbol ϕ) and it is measured either in maxwells or webers (1 weber $= 10^8$ maxwells or 10^8 lines of force). More significant than the total flux of an electromagnet is its **flux density** (*B*), or the **flux per unit** cross-sectional area. You will recall that flux density is expressed either in gauss (maxwells/cm²) or in webers/m² (one weber per square meter = 10,000 gauss). Thus if the total flux flowing through a cross-sectional area of 4 cm² is 100,000 lines (maxwells), the flux density $B = \dfrac{\phi}{A} = \dfrac{100,000}{4} = 25,000$ gauss.

(If the area is given in *square inches*, you must multiply the number of square inches by 6.452 to obtain the number of square centimeters.) Finally, let us recall that the flux density (B) induced in a certain material of **permeability** (μ) is related to the field intensity (H) by the relation

$$B = \mu \times H$$

This relation is also used to define the permeability of a medium as the ratio of flux density to field intensity, or

$$\mu = \frac{B}{H}$$

The permeability in air or in a vacuum is *unity*, and hence the flux density (in gauss) numerically equals the field intensity (in oersteds) in these media. The permeability of ferromagnetic substances is far greater than unity, reaching values as high as 100,-000 for special magnetic materials, such as *permalloy*, *hipernik*, and *perminvar*.

Strength of Electromagnet. Now let us turn back to electromagnets and their characteristics. You will recall that the field intensity (H) of a long coil (solenoid) is 4πN I/10 l, where N is the number of turns, I the current (in amps), and l the length of the coil (in cms). Let us substitute $n = \dfrac{N}{l}$ for the *number of turns per unit length* and we obtain for the field intensity

$$H = 0.4 \pi n I$$

where n is the number of turns per centimeter of length. If an iron core is inserted into the solenoid, making it an electromagnet, the flux density (B) induced in the core of the electromagnet is simply the product of the field intensity and the permeability of the iron, or

$$B = \mu \times H = \mu \times 0.4 \pi n I$$

from the relations stated before. Finally, let us multiply out the factor 0.4π and convert n from turns per centimeter to *turns per inch* by dividing by 2.54 (1 inch = 2.54 cms). Thus we obtain

$$B = \frac{0.4 \times 3.14159 \times \mu \times nI}{2.54} = \frac{1.2566}{2.54} \mu n I$$

or $B = 0.495 \mu n I$

This equation shows that the strength of an electromagnet depends only on the number of turns, the magnitude of the current and the permeability of its iron core. The product of the current (in amperes) and the number of turns (per unit length), which determines the magnetizing force (H), is known as **ampere-turns**. We may thus further simplify our formula, obtaining finally:

Flux Density of Electromagnet = 0.495 × permeability × ampere-turns (per unit length)

As we shall see presently, the only hitch in this convenient formula is that the permeability does *not* remain constant for all values of the magnetizing current because of the phenomenon of hysteresis.

EXAMPLE: What is the strength (flux density) of a 12-inch long electromagnet wound with 600 turns of wire, if the permeability of its core has a value of 2000 and a current of 8 amperes flows through the electromagnet?

Solution: The coil of the magnet has 600/12 or 50 turns per inch, and since the current is 8 amps, there are $8 \times 50 = 400$ ampere-turns per inch. Hence, the flux density

B = 0.495 × permeability × ampere-turns per inch

B = 0.495 × 2000 × 400 = 396,000 *gauss* (or 39.6 webers/m²).

Hysteresis. We have seen that magnetization of a material does *not* occur suddenly, but is a gradual process with the induced flux density slowly increasing as the "domains" jump into alignment with the external magnetizing field. The amount of magnetization flux density (B) depends on the strength of the magnetizing field (H), the process being completed when all the domains have aligned themselves with the external field. Magnetic saturation is said to occur at this point and no amount of increase in the magnetizing field (or current) can produce a further increase in magnetization. The lagging of the magnetization produced in a material behind the magnetizing force is called **hysteresis** and the amount of energy wasted (in heat) in aligning the domains is known as **hysteresis loss.** Hysteresis losses are an important factor in determining the quality of a magnetic material.

Let us look at the phenomenon of hysteresis in greater detail. Fig. 88 illustrates the **magnetization** curve of an initially unmagnetized ferromagnetic material. The abscissa of the graph shows the *intensity* of the magnetizing force H (in oersteds), which depends on the strength of the current and the number of turns (i.e., *ampere-turns*) of the elec-

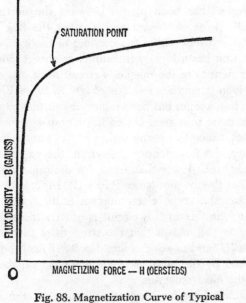

Fig. 88. Magnetization Curve of Typical
Ferromagnetic Material

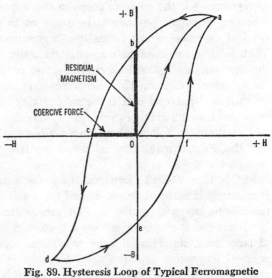

Fig. 89. Hysteresis Loop of Typical Ferromagnetic
Material

tromagnet. The ordinate of the graph shows the amount of flux density B (in gauss) *induced* in the material by the magnetizing field. The magnetization curve thus shows the variation of the flux density (B) induced in a material with corresponding changes in the magnetic intensity (H). Since it shows the dependence of B on H, the curve is sometimes simply called a **B-H curve.**

As illustrated in the magnetization curve (Fig. 88), the flux density increases rapidly at first with increasing magnetizing force, and then much more slowly as magnetic saturation is approached. Beyond the saturation point the flux density increases very little, if at all, with large increases of the magnetizing force. Since the graph is *non-linear* (i.e., *not* a straight line), the ratio of B/H or the *permeability* ($\mu = $ B/H) *does not remain constant.* Except for the lower straight-line portion of the graph, the permeability of a ferromagnetic material thus varies continuously with the magnetizing force (H) and the latter must be specified for a permeability figure to be significant. To determine the permeability of a material for a particular magnetizing force (or current), compute the ratio of B/H at the particular point of the magnetization curve, or simply measure the *slope of the curve* at that point.

The full story of magnetization is revealed by the hysteresis loop illustrated in Fig. 89. Here we have taken a sample of ferromagnetic material that was originally completely unmagnetized and have

magnetized it to saturation along curve o-a of Fig. 89. Note that this portion of the hysteresis curve is essentially identical with the magnetization curve of Fig. 88.

We now reduce the magnetizing force (H) gradually to zero and observe—with some astonishment—that the flux density (B) does *not* go back to zero along the original curve o-a, but rather follows curve a-b. It appears that the material has suddenly developed a sort of magnetic "memory" and "recalls" its previous state of magnetization. As a result, the values of the flux density obtained during demagnetization are all *larger* than those recorded during magnetization. This lag of B behind H is called **magnetic hysteresis.** It is caused by the gradual reversal and realignment of the domains during demagnetization. The amount of magnetism remaining, when the magnetizing force has been reduced to zero (segment b-o), is known as **residual magnetism** or **remanence.**

If we now reverse the direction of the magnetizing current and, hence, the *sign* of the magnetizing force H, the flux density will continue to fall along curve b-c, until for a certain negative value of the magnetizing force H the flux density reaches zero. The value of the negative magnetizing force required to demagnetize the sample completely (segment o-c) is called the **coercive force.**

Further magnetization in the negative direction establishes magnetization in the opposite direction (i.e., with north- and south-poles interchanged), along curve c-d. Magnetic saturation in the opposite direction is reached at point *d.* If H is now once again reduced to zero, B falls off along curve d-e, the residual magnetism in the opposite direction (segment e-o) being the same as in the original direction (segment o-b). With the magnetic force H again increasing, the flux density reaches zero along curve e-f, the coercive force in the opposite direction (segment f-o) being the same as in the original direction of magnetization (segment o-c). With further increases in magnetization, the flux density increases again along curve f-a and reaches saturation at point a. This completes one entire cycle of the hysteresis loop. If the magnetizing force is carried through another cycle, the hysteresis loop will continue to follow along curve a-b-c-d-e-f-a, and the original magnetization curve o-a is never repeated.

Fig. 90 illustrates two hysteresis loops for a hard ferromagnetic material (curve a) and for a soft ferromagnetic material (curve b). The magnetically hard material is characterized by a hysteresis loop of large area, signifying a large amount of stored residual magnetism and the need for a large coercive force to demagnetize the material. Such hard ferromagnetic materials (steel and various alloys) are suitable for *permanent* magnets, which must store large amounts of magnetism and resist surrounding demagnetizing forces. Curve b for soft iron, in contrast, has a small area within its hysteresis loop, signifying a small residual magnetism and a small coercive force. Such materials waste little energy in hysteresis losses and are therefore suitable for the cores of electromagnets. Since electromagnets are subject to continually reversing magnetizing fields, the hysteresis losses of their cores are of some importance. In general, the smaller the area of the hysteresis loop for a certain material, the lower are the hysteresis (heat) losses of the material.

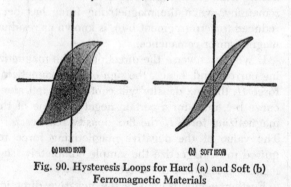

(a) HARD IRON   (b) SOFT IRON

Fig. 90. Hysteresis Loops for Hard (a) and Soft (b) Ferromagnetic Materials

Magnetic Circuits. The lines of force of magnetic flux form closed loops, as we have seen. The path that the flux lines follow is called the **magnetic circuit**. Fig. 91 illustrates some common types of magnetic circuits. In (a) of the figure the flux in a

permanent horseshoe magnet is illustrated. An iron "keeper" has been placed between the north and south poles of the magnet, so that the flux takes place entirely within the confines of the iron. Since the iron has a high permeability relative to air, the flux density in the magnetic circuit for a magnet of a given strength is relatively high. In (b) of Fig. 91, the iron keeper has been removed and the magnetic flux must now pass across the **airgap** between the poles. Since the permeability in the airgap is only unity, the flux density (B) within the gap is obviously much lower than in (a) for the same strength (H) of the magnet (since $B = \mu H$). In (c) of Fig. 91 a horseshoe type electromagnet is illustrated. The magnetic flux in this circuit is exactly the same as that in (b), except that the strength of magnetization (H) and hence the flux density (B) can be controlled by the number of ampere-turns on the coil of the electromagnet.

Ohm's Law for Magnetic Circuits. As the name implies, there is a certain similarity between electric and magnetic circuits. The primary difference is that in electric circuits the current carriers are **electrons**, while in magnetic circuits the flux is carried by the **lines of force**, which are, of course, imaginary quantities. Nevertheless, an analogy between the two types of circuits may be made. The force that produces a flow of electrons (current) in the electrical circuit is the electromotive force (emf). The force that produces the flux in a magnetic circuit is called the **magnetomotive force** (abbreviated mmf). The electric current (I) corresponds to the magnetic flux (ϕ). Finally, just as the resistance (R) opposes the flow of electric current, the **reluctance** (symbol R) opposes the magnetic flux in a magnetic circuit. The conductance (G) of an electric circuit, which indicates the ease of current flow, corresponds to the **permeability** (μ) of a magnetic circuit, the latter indicating the ease with which magnetic lines of force pass through a material.

The magnetic reluctance is defined mathematically by the relation

$$R = \frac{l}{\mu A}$$

where l is the length of the magnetic path in centimeters, μ is the permeability of the medium, and A is the cross-sectional area of the magnetic path in square centimeters. Recalling the equation for electric resistance ($R = \rho\, l/A$), we recognize that the permeability of a magnetic medium corresponds exactly to the **reciprocal of the resistivity** of a con-

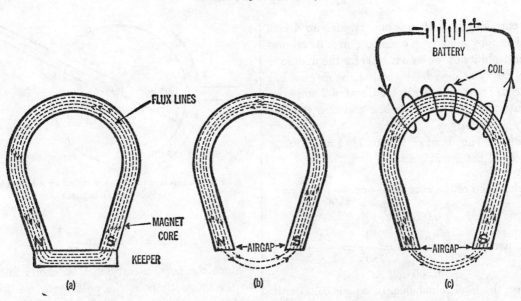

Fig. 91. Magnetic Circuits (a) Permanent Horseshoe Magnet with Keeper; (b) Permanent Magnet with Airgap; (c) Electromagnet

ductor. Reluctance does not have a specific name. The unit of reluctance is that of a magnetic circuit 1 cm in length, 1 cm² in cross section, and with unity permeability. Reluctances in series or in parallel are combined just like resistances.

EXAMPLE 1: The core length of the permanent magnet illustrated in Fig. 91 (a) is 45 cms, the length of the keeper 5 cms. The permeability of the iron used in the magnet and keeper is found from the B-H curve to be 1200 for the magnetization achieved. The cross-sectional area of magnet and keeper is 5 sq. centimeters. What is the reluctance of the entire magnetic circuit?

Solution: The total path length in iron is 45 + 5 = 50 cms. Hence, the reluctance $R = \dfrac{l}{\mu A} =$

$$\frac{50}{1200 \times 5} = 0.00833 \text{ unit.}$$

EXAMPLE 2: The magnet illustrated in Fig. 91 (b) has the same permeability and the same dimensions as that shown in (a), but the 5-cm long keeper has been left off. Compute the total reluctance of this circuit.

Solution: The reluctance of the 45-cm long iron path is $R = \dfrac{1}{\mu A} = \dfrac{45}{1200 \times 5} = 0.0075$ unit

The reluctance of the airgap ($\mu = 1$) is

$$R = \frac{1}{\mu A} = \frac{5}{1 \times 5} = 1.0000 \text{ unit}$$

Hence, the total reluctance is the sum = 1.0075

units. So we see that the addition of the small air-gap has increased the total reluctance of the circuit by a factor of more than 120.

Having drawn the analogy between electric and magnetic circuits, we can simply state **Ohm's Law for magnetic circuits: The total magnetic flux in a magnetic circuit is directly proportional to the magnetomotive force and inversely proportional to the reluctance of the circuit.** As a word equation,

$$\text{Magnetic Flux} = \frac{\text{Magnetomotive Force}}{\text{Reluctance}}$$

or in symbols ϕ (maxwells) $= \dfrac{F \text{ (mmf in gilberts)}}{R}$

Similarly, the mmf: $F = \phi \times R$

and the reluctance: $R = \dfrac{F}{\phi}$

The mmf (F) may be either expressed in *gilberts* or in *ampere-turns*, since it is directly proportional to the current (in amperes) and the number of turns of the coil of an electromagnet. Thus, an electromagnet having a winding of 2000 turns and carrying a current of 5 amperes has an mmf of $5 \times 2000 = 10,000$ *ampere-turns*. To use Ohm's Law for magnetic circuits, the mmf must be expressed in gilberts, rather than ampere-turns. To convert ampere-turns to gilberts, use the following relation:

mmf (in gilberts) $= 1.2566 \times$ ampere-turns

or mmf $= 1.2566 \, N \, I$

where N is the number of turns and I is the current in amps.

EXAMPLE: The electromagnet shown in Fig. 91 (c)

has a core length of 36 cms and an air gap 4 cms long. The core and airgap cross section is 8 cm² and the permeability of the core is 1500 for the magnetic field intensity achieved. The coil of the magnet has 500 turns of wire and carries a current of 2 amperes. Compute the total magnetic flux and also the flux density in the airgap.

Solution: The mmf = 1.2566 N I = 1.2566 × 500 × 2 = 1,256.6 maxwells.

The reluctance of the core $= \dfrac{1}{\mu \, A} = \dfrac{36}{1500 \times 8} =$ 0.003 unit

The reluctance of the airgap $= \dfrac{1}{\mu \, A} = \dfrac{4}{1 \times 8} =$ 0.500 unit

Hence, the total reluctance, R, equals 0.503 unit.

The total flux $\phi = \dfrac{F}{R} = \dfrac{1256.6}{0.503} = 2,500$ *maxwells.*

The flux density in the airgap $B = \dfrac{\phi}{A} = \dfrac{2500}{8} =$ *312.5 gauss.*

FORCE ON A CONDUCTOR (AMPERE'S LAW)

We have seen that a current-carrying conductor is surrounded by a magnetic field. We have also seen that a magnetic field exerts a force on magnetic substances within that field. The question naturally arises whether a **magnetic field exerts some force on a current-carrying conductor.** As a matter of fact, this is the case, and it is found that the interaction between the field of a current-carrying conductor and an external field actually exerts a force (of attraction or repulsion) on the conductor. The effect was discovered by the French scientist André M. Ampere, who formulated it into a quantitative relation known as **Ampere's Law.**

To understand the derivation of Ampere's law, consider a current-carrying loop, at the center of which an isolated north pole of strength m is placed (see Fig. 92). According to the relation previously given, the field intensity (*H*) at the center of such a circular loop of radius r is $H = \dfrac{2 \pi I}{10 \, r}$; where *I* is the current flowing in the loop. The magnetic pole is acted upon by a force (to the right of the page) of magnitude F = m H, as we have already described. Consequently, the force on the pole is

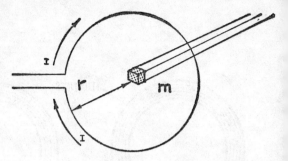

Fig. 92. Finding the Force Exerted by a Magnetic Pole on a Conductor

$$F = m \, H = m \, \frac{2 \pi I}{10 \, r} \qquad (1)$$

Since the current-carrying loop exerts this force on the magnetic pole, the **pole reacts on the loop with an equal and opposite force,** according to the **principle of action and reaction** (Newton's Third Law). The force exerted by the pole *on* the coil, therefore, is also given by

$$F = m \, \frac{2 \pi I}{10 \, r}$$

Let us multiply both numerator and denominator of this expression by *r* (which does not change its value) and rewrite, as follows:

$$F = \frac{m}{r^2} \times \frac{I}{10} \times 2 \pi r \qquad (2)$$

Equation (2) is mathematically the same as equation (1), but is more useful to us in this form.

We already know that the field intensity *H* near a pole of strength *m* is given by

$$H = \frac{m}{\mu \, r^2} \qquad (3)$$

Multiplying through by μ:

$$\mu \, H = \frac{m}{r^2}$$

But from the definition of permeability, $\mu \, H = B$ (flux density)

Hence, $\mu \, H = B = \dfrac{m}{r^2}$, and the first term of the product in equation (2), m/r^2, is equal to the flux density *B* of the field at *r* centimeters from a pole of strength *m*. Moreover, the third term of the product in eq. (2), $2 \pi r$, is simply the *length of the loop* by the well-known geometrical formula. Let us call this length *L*. Substituting these equivalent terms back in equation (2), we obtain

$$F = \frac{m}{r^2} \times \frac{I}{10} \times 2\pi r = B \times \frac{I}{10} \times L$$

or more simply, $F = \frac{BIL}{10}$ (dynes)   (4)

which is easy to remember. If B in eq. (4) is given in gauss, I in amperes, and L in centimeters, the force F is in *dynes*.

Now here is a remarkable thing. Although equation (4) has been derived specifically for the force exerted by a magnetic field on a circular, current-carrying loop, the equation turns out to be true for the force on a conductor of *any shape whatever*. This equation, which is thus generally true, is the mathematical form of **Ampere's Law**, which may be stated in words as follows: **Any current-carrying conductor located in a magnetic field at right angles to the lines of force will be pushed by a force that is directly proportional to the flux density, the current and the length of the conductor.**

Right-Hand Rule for Force on a Conductor (Motor Action). If you insert a straight, current-carrying conductor at right angles to the flux of a magnetic field, the directions of the current I, the flux density B and the force on the wire F will be mutually perpendicular, as illustrated in Fig. 93a. A simple rule, known as the **right-hand motor rule**, is used to determine the *direction* of the force on the conductor: **Extend the thumb, index finger and middle finger of the right hand at right angles to each other, so that the index finger points in the direction of the magnetic flux (north to south), and the center finger points in the direction of the elec-**tron current ($-$ to $+$); the outstretched thumb will then indicate the direction of the force exerted on the conductor and hence the direction in which it tends to move. The right-hand motor rule is illustrated in Fig. 93b.

Composite Magnetic Field of a Magnet and Current-Carrying Conductor. The interaction between the magnetic field of a magnet and that of a current-carrying conductor produces a force on the conductor, which is the basis of operation of many devices, such as electric motors and galvanometers. The interaction between the fields when a current-

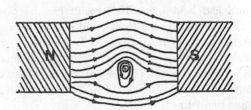

Fig. 94. Composite Magnetic Field of Magnet and Current-Carrying Wire

carrying wire is inserted between the poles of a magnet is illustrated in Fig. 94.

The cross section of the wire in Fig. 94 is shown by the heavy circle and the direction of the current (out of the page towards you) is indicated by the dot $\odot$ within the circle. By the left-hand rule for a conductor, given earlier in this chapter, the direction of the field produced by the current in the wire will be clockwise, as is indicated by the arrows. The magnetic flux *above* the wire is thus *aided* by the

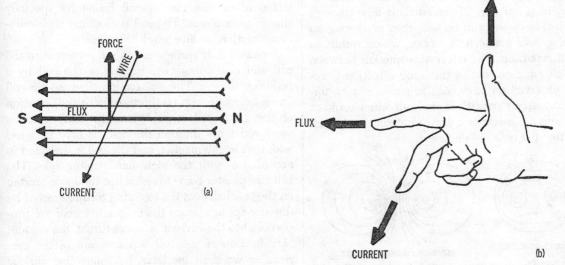

Fig. 93. Direction of Force on a Conductor (a) and Right-Hand Motor Rule (b)

field surrounding the wire, while the flux *below* the wire is *opposed* by the field of the wire. As a result, the lines of flux are *concentrated above the wire* and they are *weakened below the wire*. Thus, there is a relatively strong force above the wire and a relatively weak force below the wire. The stronger force prevails, of course, and the *wire will be pushed downward into the weaker field*. You might also look at the action by thinking of the lines of force as elastic rubber bands, which tend to straighten themselves out and become as short as possible. Again, the many stretched lines above the conductor will result in pushing the conductor *downward* toward the fewer and less bent lines. This result can be obtained more simply by the right-hand motor rule.

EXAMPLE: If the wire in Fig. 94 is 10 cms long and carries a current of 6 amps, and if the flux density between the poles is 10,000 gauss, what is the magnitude and direction of the force on the wire? (See Fig. 94.)

Solution: By equation (4), the force on the wire is

$$F = \frac{B I L}{10} = \frac{10,000 \times 6 \times 10}{10} = 60,000 \text{ dynes.}$$

By the right-hand rule, the direction of the force is *downward*.

Force Between Parallel Conductors. Since every current-carrying conductor is surrounded by a magnetic field, the fields of *two* current-carrying wires will also interact with each other. Fig. 95 illustrates the magnetic fields about two parallel conductors. In (*a*) of the figure, the electron current in each wire flows in the *same direction* (out of the page) and hence the magnetic fields *around* the wires aid each other, as shown by the arrows, and draw them together. In (*b*) of Fig. 95 the currents flow through the wires in *opposite* directions, thus producing an *opposing field between* the wires, which results in mutual repulsion. The effect of attraction between wires carrying current in the same direction is actually observed on large coils in power plants; under short-circuit conditions the individual turns of these coils are sometimes drawn together with such force that the coils are damaged.

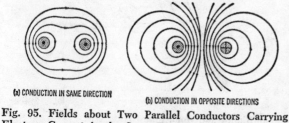

(a) CONDUCTION IN SAME DIRECTION

(b) CONDUCTION IN OPPOSITE DIRECTIONS

Fig. 95. Fields about Two Parallel Conductors Carrying Electron Current in the Same Direction (a) and Carrying Currents in Opposite Directions (b)

The Galvanometer. An important application of the force experienced by a conductor in a magnetic field is the **moving-coil** or **d'Arsonval galvanometer** (named after the French physicist ARSENE D'ARSONVAL). Based on the reaction of a current-carrying coil suspended in the field of a permanent magnet, the d'Arsonval galvanometer is specifically designed for the detection of extremely small currents; its movement is also the heart of most present-day current and voltage measuring instruments. The basic d'Arsonval movement is illustrated in Fig. 96. The

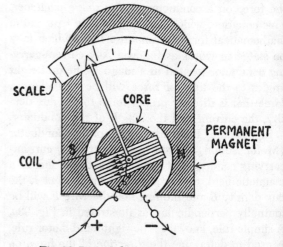

Fig. 96. Basic Galvanometer Movement for Current Measuring

moving coil, consisting of some 20 turns of insulated copper wire, is wound around a light aluminum frame and is free to turn about a soft-iron core. The external magnetic field is provided by a strong, horseshoe-type permanent magnet and is concentrated about the moving-coil frame by specially shaped pole pieces. The coil is mounted with minimum friction on fine jewel bearings.

A pair of hair springs conduct the current to the coil and also oppose the turning of the coil by a restoring torque. The pointer attached to the coil frame normally rests on the zero (left-hand) position of the graduated scale. When the current to be measured flows through the coil, its field interacts with that of the magnet, and the coil is deflected in accordance with the right-hand motor rule. The coil and pointer come to rest when the force exerted on the coil balances the restoring torque exerted by the hair springs. Since the magnetic forces are proportional to the current (ampere-turns), the amount of deflection of the coil is a measure of the current. As we shall see later, by connecting various resistors in series or in parallel with the galvanometer movement, the scale of the meter can be calibrated directly in microamps, milliamps, volts, or even in ohms.

EXPERIMENT 18: Let us construct a simple current-indicating instrument having a fixed coil and a moving compass needle (Fig. 97). Wind about 25 turns of bell wire around a bottle, leaving a foot of straight wire on each end. Slip the coil off the bottle, tape the turns together and mount the coil in an upright position on a wooden base, as illustrated in Fig. 97. Now place a pocket compass

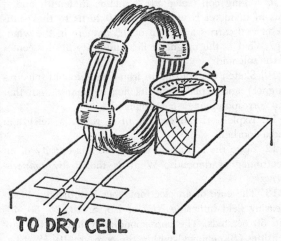

TO DRY CELL

Fig. 97. Construction of Simple Current Indicator (Experiment 18)

next to and near the center of the coil (on a wood block) and rotate the entire arrangement to a north-south direction, so that the coil lines up parallel with the compass needle.

To show that the meter works, connect the coil to a dry cell and check the direction of compass needle deflection with the left-hand rule for coils. You can make a rough calibration of the scale of the compass by connecting the meter to *known* currents, calculated in accordance with Ohm's Law. (Don't forget to include the resistance of the coil in these calculations.)

The Direct-Current Motor. If the current-carrying coil of the galvanometer we have just described could be made to turn continuously, it would provide a *source of mechanical energy from a supply of electric energy.* Any rotating device that converts electrical into mechanical energy is called an **electric motor.** The galvanometer cannot be made to turn more than half a revolution at most (no matter how great the current), since the moving coil comes to rest as soon as its lines of force line up with the external magnetic field of the permanent magnet. To achieve continuous rotation we need some sort of device that reverses the relative polarity of the

two fields every half turn. Such a polarity-reversing device is called a **commutator,** and we shall become acquainted with its action presently.

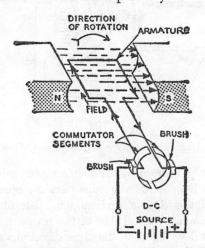

Fig. 98. Simplified Construction of Electric (D-C) Motor

A simplified model of a direct-current motor is shown in Fig. 98. Its essential elements are the **armature,** the **magnetic field,** the **commutator,** and the **brushes.** The armature is usually a cylinder of soft steel about which a number of turns of copper wire are wound, constituting the **armature winding.** For simplicity, Fig. 98 shows only a single loop of wire serving as armature. The external field in which the armature rotates is usually provided by a **multi-pole electromagnet,** but the basic action is the same for the field of a two-pole permanent magnet, shown in Fig. 98. The commutator is attached to the shaft of the armature and is essentially a **reversing switch.** It consists of as many ring-shaped segments as there are poles in the magnet (in this case two). Current is passed to and from the armature through graphite (carbon) **brushes,** which slide on the cylindrical commutator segments. A battery or generator serves as direct-current source.

Assume that an electron current initially flows through the armature in the direction indicated by the arrows. By applying the right-hand motor rule you can verify that the *left* conductor of the armature (nearest the N-pole) undergoes an *upward force,* while the *right* conductor is subjected to a *downward force.* As a result the armature turns clockwise until the plane of the loop is *vertical.* Without the commutator it could not turn further, since beyond this point the left and right conductors (and hence the current directions) are interchanged and the forces would be reversed. At that very moment, the current through the armature is

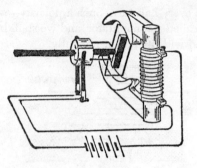

BATTERY

Fig. 99. Demonstration-Type Direct-Current Motor

automatically reversed by the switching of the connections as the commutator gaps pass the brushes. Because of this *double* reversal (i.e., that of the armature conductors and that of the commutator), the current flows again in the same directions relative to the field and the armature coil makes another half turn. At that point the commutator switches the armature connections once more, with another half turn of the loop resulting. The upshot of the action is that the armature coil turns continuously in *one direction*. The direction can be reversed by either switching around the battery connections or by reversing the polarity of the magnetic field. Useful work can be obtained from the electric motor by attaching some load to its output shaft. In practice, there are many other refinements necessary to obtain an efficiently operating motor. A demonstration-type motor, using an electromagnet for the field and a single-turn armature, is illustrated in Fig. 99.

Practice Exercise No. 9

1. A horizontal, current-carrying wire is surrounded by circular lines of force in a clockwise direction (like a corkscrew). What is the direction of electron flow?

2. Two long, parallel wires 10 inches apart carry currents of 20 amps each, in opposite directions. What is the magnetic field intensity at a point midway between the wires?

3. A pair of str   parallel wires carry equal currents. At what   the resulting magnetic field about the wires   vhen the currents flow in the *same* direction? Can you find a point where the resultant field is zero, when the currents flow in opposite directions?

4. What is the current flowing through a circular wire loop of 20 cms diameter if the field intensity at the center of the loop is 5 oersteds?

5. State the electromagnetic definition of the ampere.

6. A current of 2 amps flows through a circular, flat coil of 50 turns and 12 cms in diameter. What is the field intensity at the center of the coil?

7. What is the field intensity inside a ring solenoid of 300 turns per cm length and carrying a current of 2.5 amps?

8. Draw a diagram of an air-core solenoid powered by a dry cell. Show the direction of the current, the resulting polarity of the solenoid and sketch the magnetic field inside and outside of the solenoid. State the rule used to determine the direction of the field for this type of coil.

9. A long coil (solenoid) 3 meters in length and 4 cms in diameter is wound with 10 turns to the centimeter and carries a current of 0.5 ampere in the winding. What is the magnetic field intensity at the center of the solenoid?

10. Sketch the field of an iron-core solenoid (electromagnet) and explain why its flux is greater than that of a corresponding air-core solenoid.

11. Explain the operation of a relay, a telegraph and a doorbell.

12. State the factors on which the flux density of an electromagnet depends. What is meant by *ampere-turns*?

13. The core of an electromagnet is subjected to increasing field intensities (H) of 2.5, 5, 10, 20, 30, 40, and 50 oersteds. The corresponding values of the flux densities (B) achieved within the core are 10, 12, 13.4, 14.4, 14.9, 15.3 and 15.6 kilogauss. Is the permeability of the iron core constant, increasing or decreasing? To what do you attribute its behavior?

14. Draw hysteresis loops for soft and hard ferromagnetic materials and contrast their residual magnetism, their coercive forces, and the total hysteresis losses. Which type of material is best suited for an electromagnet, which for a permanent magnet? Why?

15. State "Ohm's Law" for magnetic circuits and define each of the quantities.

16. State the factors on which the reluctance of a material depends upon and write the formula. If reluctances in series and in parallel are added like resistances, can you derive a formula for reluctances in parallel?

17. If the magnetomotive force is expressed in gilberts, what factor would you apply to convert the mmf to ampere-turns?

18. If the mmf applied to an electromagnet is doubled, while the reluctance of its magnetic circuit is reduced to one-half, how does this affect the total flux? (Assume μ is constant.)

19. Two parallel, straight wires, separated by a distance of 8 cms, carry currents of 40 and 50 amps, respectively. What force does either wire exert upon the other per cm of length?

20. Explain the action of a moving-coil galvanometer and that of a direct-current motor. Draw a sketch of each, showing the essential elements. State the rule

used to determine the direction of motion in each of these devices.

SUMMARY

An **electric current** is always surrounded by a **magnetic field**. The **field intensity** in a straight conductor varies *directly* with the *strength* of the current through the conductor and *inversely* with the *distance* between the wire and the point of measurement.

Left-Hand Rule for Conductors: With the thumb pointing in the direction of electron flow, the fingers (of the left hand) wrapped around the wire will point along the lines of force.

Field intensity of straight wire:

$$H = \frac{2\,I}{10\,r} \text{ oersteds.}$$

Field intensity at center of **circular loop**:

$$H = \frac{2\,\pi\,I}{10\,r} \text{ oersteds}$$

Field intensity at center of **flat coil** of N turns:

$$H = \frac{2\,\pi\,N\,I}{10\,r}$$

Field intensity near center of **solenoid** (length l):

$$H = \frac{4\,\pi\,N\,I}{10\,l}$$

Left-Hand Rule for Long Coils (Solenoids): Wrap fingers of left hand around coil in direction of electron flow; the thumb will point in the direction of the north pole of the field.

An **electromagnet** is a coil of wire wound around a soft-iron core. The iron core multiplies the flux density for the same magnetizing force compared with an air core coil.

A **relay** is a **remote-control switch** that permits control of a heavy current by a weak current. Its heart is an electromagnet. Telegraphs, teletype machines, buzzers and electric doorbells are also based on the operation of electromagnets.

The **strength of an electromagnet** depends on the **number of turns** of its winding (per unit length), the **magnitude of the current**, and the **permeability** of its iron core. The product of the current (amperes) and the number of turns, called **ampereturns**, determines the magnitude of the **magnetizing force (H).**

Magnetic saturation occurs when most of the domains in a magnetic material are aligned with the external magnetizing field. Beyond saturation further increases in magnetization (flux density B) cannot be attained with increasing magnetizing force and the permeability of the material drops off.

The lagging of the flux density (B) produced in a material behind the magnetizing force (H) is called magnetic **hysteresis**. The amount of magnetism remaining in a material after the (de)magnetizing force has been reduced to zero is called **residual magnetism**; the value of the negative magnetizing force required to demagnetize a sample completely is known as **coercive force**.

The **area enclosed by the hysteresis loop** is a measure of the energy wasted in heat to magnetize and demagnetize a material; the heat energy wasted is known as **hysteresis loss**. Hard iron, suitable for permanent magnets, has a large-area hysteresis loop; soft iron, suitable for electromagnets, has a small-area hysteresis loop.

Ohm's Law for Magnetic Circuits: The total magnetic flux in a magnetic circuit is directly proportional to the magnetomotive force (mmf) and inversely proportional to the reluctance. $(\phi = \frac{F}{R})$.

Ampere's Law for the Force on a Conductor: Any current-carrying conductor located at right angles to the lines of force of a magnetic field will be pushed by a force that is directly proportional to the flux density, the current, and the length of the conductor. $(F = B\,I\,L/10.)$

Right-Hand Motor Rule:

Fore (Index) Finger   = *Flux*
Center (Middle) Finger = *Current*
Thumb   = *Motion* or *Force*

Moving-coil (d'Arsonval) **galvanometers** depend for their operation on the interaction of the field about a current-carrying coil and that of a permanent magnet.

Electric motors convert electrical energy into mechanical energy. A **direct-current motor** consists essentially of a current-carrying **armature**, a magnetic field produced by a permanent or electromagnet, a **commutator** for reversing the current direction in the armature every half-turn, and graphite brushes to feed current to the armature winding through the commutator segments. A motor operates because of the interaction of the magnetic field of the armature coil and that of the field magnet, but in contrast to the galvanometer, continuous rotation is made possible by the automatic current-reversing action of the commutator.

CHAPTER TEN

INDUCED ELECTROMOTIVE FORCE

The astounding development of present-day commercial electric power and technology started about 125 years ago with the discovery of the principle of induction. After Oersted and others had shown that magnetism was associated with and could be produced by electricity, many scientists started to look for the reverse effect—the production of electricity from magnetism. MICHAEL FARADAY in particular initiated a series of experiments which after seven years of painstaking work culminated in the discovery of **electromagnetic induction**, the principle of the **generator**. In the United States the scientist JOSEPH HENRY (1797-1878) independently discovered the induction or generator principle, which led to the commercial development of alternating currents, transformers, and a host of other devices associated with modern electricity. Although Faraday reported discoveries first (in 1831), both Faraday and Henry should be credited with this key discovery.

Faraday's Experiments. In one of the simplest and most basic of Faraday's experiments, he connected a coil directly to a current meter (galvanometer) and pushed a bar magnet into and out of the coil, as illustrated in Fig. 100. He found that a momentary current was registered on the meter, whenever he moved one pole of the magnet quickly toward the coil (or the coil toward the magnet). When he jerked the magnet away from the coil, there was again a brief current "kick," but this time in the *opposite* direction, as registered by the re-

verse meter deflection. No current was observed as long as the magnet and the coil were held still in any position.

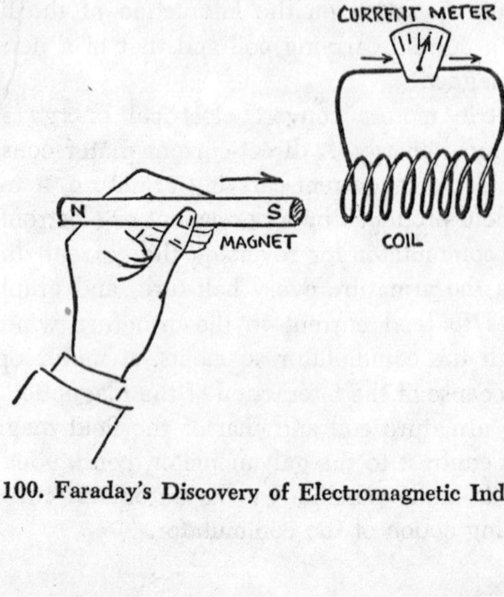

Faraday made some further observations by changing various factors involved in the simple experiment. He found that the magnitude of the momentary current registered by the meter depended on the *speed* of moving the magnet toward or away from the coil, the *strength* of the bar magnet and the *number of turns* of wire on the coil. Whenever there was relative motion between the magnet and the coil, a momentary current was registered by the meter. The *direction* of the current depended on whether the motion of the magnet was toward or away from the coil and also on which pole of the magnet was pushed into the coil. When the north and south pole of the magnet were interchanged the meter would register a current pulse in the opposite direction. Moreover, Faraday found that the bar magnet could be replaced by an **electromagnet** with the same results. *Increasing* the current in the electromagnet had the same effect as moving it *toward* the coil, and *decreasing* the current had the same effect as moving the magnet away from the coil. Thus, *no motion at all* was required with an electromagnet to produce the phenomenon of induction.

In another of Faraday's experiments, he wound two coils, carefully insulated them from each other, and arranged them on the same axis in close proximity to each other (see Fig. 101). (He actually wound both coils on a wooden cylinder.) He then

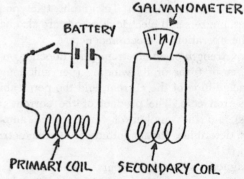

Fig. 100. Faraday's Discovery of Electromagnetic Induction

Fig. 101. Faraday's Experiment Showing Induction
Between Two Coils

connected one coil (the **primary**) through a switch to a battery and the other coil (the **secondary**) to a galvanometer. Whenever he closed the switch in the **primary circuit**, Faraday noted a momentary current "kick" on the meter of the entirely separate **secondary circuit**. Nothing further happened, once a steady current was established in the primary circuit. If he now *opened* the switch in the primary circuit, he observed another kick of current on the meter of the secondary circuit, but in the *opposite* direction. Again, nothing further happened after the brief current pulse during the opening of the circuit. Faraday then varied the amount of current in the primary coil and found again that the meter was deflected with every *change* in the primary current, the needle being deflected in one direction for an *increase* in the current and in the opposite direction for a decrease. He finally concluded that the effect could be observed whenever the primary current was *changing* (such as in opening or closing the circuit) but that nothing happened when the current remained unchanged.

After various experiments, Faraday produced his "new electrical machine," now known as Faraday's **disk dynamo**, which is the forerunner of the modern generator. As illustrated in Fig. 102, this machine consists of a 1-ft copper disk, mounted so that it could turn freely between the poles of a strong horseshoe magnet. Two copper brushes bore against the shaft and the circumference of the disk, respectively. When the disk was rotated in the magnetic field by means of the handle, a feeble but *continuous emf* was generated between the brushes, as indicated by the current through the meter. The direction of rotation determined the direction of the emf and hence of the current through the meter.

Faraday's Laws. Faraday thought of some simple, graphic explanations for the phenomenon of induction which are still in use today. Obviously, if

a current is flowing in some circuit due to the phenomenon of induction, an **electromotive force** must have given rise to it. The basic question is how an emf can be induced in a conductor which then gives rise to a current in a **closed circuit**. Faraday was able to generalize from all his experiments that an **emf was induced** in a loop of wire located in a magnetic field, whenever the number of lines of force (or **flux**) passing through the loop was *changing*. If the loop was *closed*, the induced emf would give rise to a current through the circuit. The flux *linking* the loop could be *expanding* or *collapsing*, such as would happen, for example, if the primary circuit of Fig. 101 is *closed* or *opened*. The flux linking the loop might merely *vary in strength*, such as when a magnet is brought near to or moved away from it (see Fig. 103). All that is necessary to produce an emf is a change in the total flux linking the loop. Nothing happens as long as the flux through the loop remains the same.

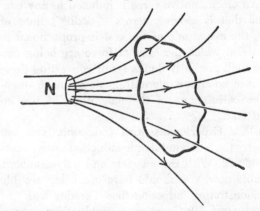

Fig. 103. Magnetic Flux Linking a Loop of Wire

In some experiments Faraday thought it more convenient to visualize the process of induction as the "cutting" of the lines of force of a magnet or electromagnet by a wire or a coil. An emf would be induced in the wire or coil whenever it was *moving across lines of force* established by a magnet or electromagnet. For example, when the wire in Fig. 104 is moved across one pole of a stationary magnet, it will *cut across the lines of force*, thus inducing an emf between its ends; this emf can be measured with a voltmeter. If the wire is connected into a closed circuit, a current will flow.

These two different viewpoints of visualizing the fundamental fact of electromagnetic induction can be summarized by two **laws of induction**, as follows:

1. An electromotive force is induced in a coil of

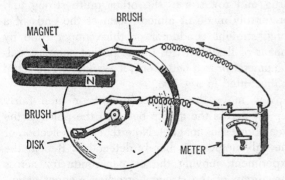

Fig. 102. Faraday's Disk Dynamo for Producing a
Continuous Emf by Electromagnetic Induction

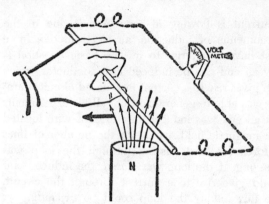

Fig. 104. When a Wire is Moved Across the Lines of
Force of a Magnet, an Emf is Induced Between
the Ends of the Wire

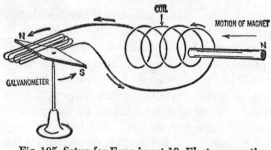

Fig. 105. Setup for Experiment 19: Electromagnetic
Induction

wire whenever the number of lines of force (magnetic flux) linking the coil is changing; the magnitude of the induced emf is proportional to the rate at which the number of lines of force through the coil are changing.

2. An electromotive force is induced in any conductor that is moving across ("cutting") lines of force; the magnitude of the emf is proportional to the rate at which the lines of force are being cut.

We shall presently find ways of converting these laws into equations, permitting us to make quantitative calculations of the magnitude of the induced emf.

Further Experiments with Induction. You can easily try the following simple induction experiments for yourself. With the magnets and galvanometers available now, you should have much less trouble in demonstrating induction than Faraday had.

EXPERIMENT 19: Construct a simple galvanometer by winding several loops of wire around a suspended compass needle (Fig. 105) or use the current indicator constructed for experiment 18 (Fig. 97). (If you have a sensitive zero-center galvanometer available, use it instead.) Rotate the home-made galvanometer in the north-south direction so that the coil of wire is parallel with the compass needle. Connect the free ends of the galvanometer coil to another coil of wire, constructed initially of three or four turns of heavy copper wire. Obtain a fairly strong (alnico) bar magnet.

Now thrust one pole of the bar magnet *into* the coil and note the *direction* in which the compass needle deflects. Next jerk the bar magnet rapidly *away* from the coil and observe that the compass needle deflects in the *opposite* direction. Now interchange the north and south pole of the bar magnet and repeat the experiment. Note that the compass

needle deflections are again reversed with respect to the deflections obtained for the initial polarity. We conclude that the *direction* of the induced emf depends on the direction of the *magnetic field lines* and on the *direction of motion*.

Repeat the experiment by moving the bar magnet at first slowly and then rapidly into and out of the coil. Note that the *magnitude* of the needle deflection and hence that of the induced emf is in direct proportion to the *speed of motion* across the field.

EXPERIMENT 20 (*See Fig. 105*): Using the same setup as for experiment 19 (Fig. 105), make the following experiment. First reduce the number of turns of the induction coil to a *single loop*. Repeat the previous experiment, moving the bar magnet alternately into and out of the loop. Note that the deflection of the galvanometer is barely detectable, even at high speeds of motion. Now *increase* the number of turns of the induction coil first to three, then six and finally to 10 or more turns. Attempt to move the bar magnet into and out of the coil with approximately the *same speed* in each case. Note that the *magnitude* of the needle deflection, and hence that of the induced emf, is roughly **proportional to the number of turns of the coil.** (Save the setup for the experiment with Lenz's Law.)

EXPERIMENT 21: Obtain two horseshoe magnets, one relatively weak, such as the variety carried by drug and toy stores, the other quite strong and preferably made of alnico. Connect the ends of a long, straight conductor or thin copper tube by means of flexible wires to the pocket compass galvanometer or to a better one, if available. The setup is illustrated in Fig. 106.

Now move the long conductor or tube fairly rapidly across the airgap between the poles of the weak horseshoe magnet. Note that the deflection of the galvanometer is barely detectable. Repeat the experiment, moving the copper conductor across the airgap of the strong horseshoe magnet at approximately the same speed as before. Note that

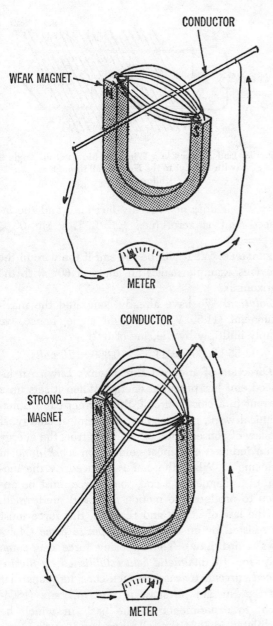

CONDUCTOR

WEAK MAGNET

METER

CONDUCTOR

STRONG
MAGNET

METER

Fig. 106. The Magnitude of the Induced Emf Varies
Directly with the Strength of the Magnetic Field
(Experiment 21)

laws of induction the magnitude of the induced emf is proportional to the rate of change of magnetic flux, or equivalently, to the rate at which lines of force are being cut. Now it has been found experimentally (and can be shown theoretically) that an *emf of 1 volt is induced in a conductor* or single loop, when it *cuts magnetic lines of force at the rate of 100,000,000 or 10^8 per second*. Equivalently, *1 volt is induced* in a single loop when the flux linking the loop changes at the rate of 10^8 lines (or 1 weber) per second. We may therefore state for the emf induced in a conductor or loop:

Induced Emf E (volts) =

$$\frac{\text{Rate of Change of Flux (maxwells)}}{10^8}$$

or E (volts) = Rate of Change of Flux (webers)
Symbolically, this may be written

$$E = \frac{\Delta \phi}{\Delta t} \times 10^{-8} \text{ volts}$$

where $\Delta \phi$ is the *change of flux* (in maxwells or lines of force) occurring in a time interval Δt. For a coil of N turns, the induced emf is simply multiplied by the number of turns linking the flux, and we obtain:

Induced Emf in Coil of N turns:

$$E = N \frac{\Delta \phi \text{ (maxwells)}}{\Delta t} \times 10^{-8} \text{ volts}$$

$$\text{or} \quad E = N \frac{\Delta \phi \text{ (webers)}}{\Delta t}$$

EXAMPLE: A coil composed of 50 turns of wire links 50,000 lines of force (maxwells). If this flux collapses in 1/100 second, what is the emf induced in the coil?

Solution: $E = N \dfrac{\Delta \phi}{\Delta t} \times 10^{-8} =$

$$\frac{50 \times 50,000}{0.01} \times 10^{-8} = 2.5 \text{ volts}.$$

For a conductor that moves across and cuts lines of force, a more convenient expression can be obtained from the second law of induction, which is equivalent to the one above. Fig. 107 illustrates a wire of length L, which moves with velocity v on a pair of metallic rails at *right angles* to a uniform magnetic field, directed into the page. The emf induced into the wire is picked off the rails, which act as brushes.

According to the second law of induction, the magnitude of the emf induced in the wire in this

the deflection of the galvanometer needle is now far greater than for the weak magnet. We conclude that the **magnitude of the emf induced depends on the number of lines of force cut per second.** In the strong magnet with a greater flux, *more* lines of force are being cut for the same speed of movement. More lines may also be cut, with a consequent increase in the induced emf, by moving the conductor *more rapidly* across the magnetic field.

Magnitude of Induced Emf. According to the

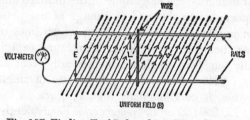

Fig. 107. Finding Emf Induced in a Wire that Moves at
Right Angles to a Uniform Magnetic Field

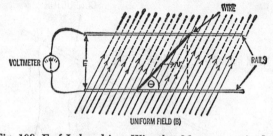

Fig. 108. Emf Induced in a Wire that Moves at an Angle Θ
(with Respect to the Horizontal) through a
Uniform Field

arrangement is proportional to the *rate* at which flux is being cut. This rate of flux cutting evidently depends on three factors: 1.) the amount of flux or the flux density B (in gauss); 2.) the *rate* at which the wire cuts across the field or the velocity *v* (in cm/sec.); and 3.) the length (*L*) of the wire (in cm) in which the emf is being induced. We also know that 10^8 lines of force must be cut each second for each volt being induced in the wire. Putting all these factors together, we obtain for the emf induced in the wire of Fig. 107

$$E = \frac{B \times L \times v}{10^8} \text{ volts}$$

or more simply $E = B \, L \, v \times 10^{-8}$ volts

EXAMPLE: A wire 33 cms (13 in.) long is moved at right angles across a magnetic field of 50,000 gauss at a speed of 7 meters per second (23 feet/second). Compute the emf induced between the ends of the wire.

Solution: $E = B \, L \, v \times 10^{-8} = 50{,}000 \times 33 \times 700 \times 10^{-8} = 11.55$ volts.

As mentioned before, this formula applies only to a wire that is moving at *right angles* or perpendicularly to the magnetic field. If the wire is moving at some angle (θ) with respect to the horizontal (the rails), as illustrated in Fig. 108, the emf induced is proportional to the *vertical projection* of the wire; that is, the *equivalent* portion of the wire that moves *at right angles* to the lines of force. You may remember from trigonometry that such a vertical projection is expressed by the *sine of the angle* (θ) between the line (wire) and the horizontal. Thus, we have to modify our previous expression by the sine of θ ($\sin \theta$) to take into account the situation depicted in Fig. 108. Hence, for any angle θ,

$$E = B \, L \, v \sin \theta \times 10^{-8} \text{ volts}$$

When the wire is *vertical* or *perpendicular* to the field ($\theta = 90°$), it will cut the *maximum* number of lines of force, and since $\sin 90°$ is 1, the expression reduces to the one obtained previously. In contrast, when the wire is *horizontal* or *parallel* to the rails

($\theta = 0°$), it does not cut any flux lines and the induced emf is zero. (i.e., $E = B \, L \, v \sin 0° \times 10^{-8} = 0$.)

EXAMPLE: What will be the emf if the wire in the previous example makes an angle of 60° with the horizontal?

Solution: We have already computed the maximum emf (11.55 V) for a right angle; hence, we simply multiply by the sine of 60°:

$$11.55 \sin 60° = 11.55 \times 0.866 = 10 \text{ volts.}$$

Direction of Induced Emf (Lenz's Law). An induced emf has potential energy. When it sets up a current in a closed circuit, this current can do mechanical work, produce heat or chemical energy. If the law of conservation of energy is true, the energy of an induced emf must come from *work done* in inducing it. When the emf is induced by the motion of a magnet or a coil, some *force* must be applied to produce the motion and work performed. By the law of action and reaction, this force must be resisted by an *equal and opposite force* (Newton's Third Law). The opposing force can come only from the *magnetic field established by the induced current*. It was just such considerations that led the Russian physicist H. F. EMIL LENZ (1804-1865) to experiments on the basis of which he postulated in 1834 the following law (Lenz's Law): A current set up by an emf induced due to the motion of a (closed-circuit) conductor will be in such a direction that its magnetic field will oppose the motion causing the emf. The induced emf will have the proper polarity to cause electron flow from − to + (or conventional current from + to −) in the direction postulated by Lenz's Law.

EXPERIMENT 22: To illustrate Lenz's Law, let us repeat experiment 19 (see Fig. 105), but this time we shall try to predict the *direction* of the induced (electron) current. Assume that the S-pole of the bar magnet is moved towards the coil, as illustrated in Fig. 105. If this action is to be *opposed* by the magnetic field of the induced current, in accordance with Lenz's Law, the side of the coil facing the magnet must also have **south polarity;** only then will the lines of force emanating from the coil oppose those of the bar magnet. Let us use the left-hand rule for coils to determine the current direction for this polarity. Wrapping the fingers of the left hand around the coil such that the thumb points toward the end of the coil *away* from the magnet (the north pole), we see that the fingers must grasp the coil counterclockwise and, hence, the direction of the electron current is *counterclockwise* (viewed from the magnet), as indicated in Fig. 105.

We can also determine the direction of the compass needle deflection by the left-hand rule for coils. By the rule, the galvanometer coil will have a north pole at its left and a south pole at its right end (facing the induction coil). Since the flux outside the coil is from north to south, the flux *inside* the coil will be from *south to north.* The compass needle will deflect so that it lines up with the direction of the field inside the coil. The north pole of the needle, consequently, is deflected *counterclockwise* (to the left), as indicated in Fig. 105. Repeat the experiment several times, moving the magnet away from the coil, reversing south and north pole, etc. and try to predict in each case the direction of the induced current and that of the compass needle deflection.

Left-Hand Rule for Generator Action. As you found out in the last experiment, Lenz's Law, though fundamental, is somewhat difficult to use, as it involves a number of detailed considerations. A more convenient rule, known as the **left-hand rule for generator action** (Fig. 109), states: **Extend the thumb, fore (index) finger and center finger of** the left hand at right angles to each other, so that the forefinger points in the direction of the flux and the thumb points in the direction of the motion; the center finger will then point in the direction of the induced (electron) current. (For *conventional* current, you must use the *right hand to determine the direction of the current and the emf.*)

You can easily verify that this rule gives the correct current direction in Fig. 105. However, some caution is necessary in applying the rule. The "motion" referred to in the rule applies to the *motion of the conductor in which the emf is induced.* You must therefore think of the coil (in Fig. 105) being moved *toward the magnet* (to the right of the page), rather than the magnet being moved into the coil (toward the left of the page). Moreover, you will obtain a current direction *toward you* for the upper portion of the coil (above the magnet) and a direction *away from you* for the lower coil portion, since the flux is *down* (from north to south) above the magnet and *up* below the magnet. This is, of course, correct and shows that the current is *counterclockwise,* as indicated.

INDUCING AN EMF IN A ROTATING COIL: THE GENERATOR

We have seen that a continuous emf is induced in a conductor that is moved at a certain speed across a magnetic field. Since you might get tired running with a wire across a magnetic field, an easier way of arranging the generation of such a continuous emf is to rotate a coil between the poles of a magnet (or electromagnet) so that the conductors cut across the lines of force. Such an arrangement is illustrated in Fig. 110 and it is the basic principle of operation of all types of **electric generators.** The type shown in the figure is an alternating-current generator, and as you can see, it is almost indistinguishable from the electric motor illustrated in Fig. 98. As in the motor, the essential parts of an electric generator are an **armature coil,** a **magnetic field** in which the coil can be rotated, and some **means of connecting** the rotating coil to an external circuit. (In the a-c generator, the brushes rotate on continuous **slip rings,** rather than a commutator, as we shall see.) The essential difference between motors and generators lies in their use. In the motor, current from an external source of power is passed through the armature coil, which then rotates in the magnetic field, doing mechanical work in the process. In the generator, a shaft at-

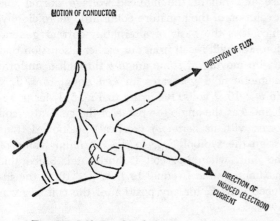

Fig. 109. Left-Hand Rule for Generator Action

tached to the armature is rotated by some *mechanical* means, such as a gasoline engine or an electric motor. The rotation of the armature coil in the magnetic field then *converts this mechanical work into electric energy*, which may be tapped off the brushes and conducted to an outside circuit. In principle, the same machine can be used *both* as motor and as generator. (In practice some adjustments are needed.)

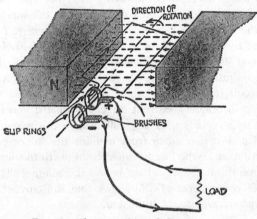

Fig. 110. Elements of Simple A-C Generator

Assume that the armature coil rotates *counterclockwise* in the magnetic field (usually provided by an electromagnet rather than a permanent one) and is initially in the horizontal position illustrated in Fig. 110. In this position the *plane* of the coil is parallel to the magnetic field and each of the two long conductors of the coil cuts the lines of force *at right angles* (perpendicular to the field). As we have seen before (Fig. 107) the maximum number of lines of force are being cut in this position and, hence, the emf induced in the coil is a *maximum*.

The ends of the coil are connected to separate **slip rings** against which individual brushes bear. If the brushes are connected to some load (resistance) in an external circuit, a current will flow through this circuit, leaving the generator at the brush marked — and returning to it through the brush marked +. (This is the direction of *electron flow;* conventional current flows in the *opposite* direction.) We can easily verify the direction of the induced current with the left-hand generator rule. Application of the rule shows that in the right-hand conductor of the loop (facing the S-pole), the induced current will flow in a direction *out of the page* (towards you), while in the left-hand conductor the current flows *into the page* (away from you). The current thus leaves at the brush marked

(—) and returns through the brush marked (+), as indicated.

The *magnitude* of the emf induced in *each* of the two long conductors is equal to B L v × 10⁻⁸ volts for the position shown, you will recall. Since the emf's induced in the two conductors *aid* each other, the total emf induced in the armature coil will be 2 B L v × 10⁻⁸ volts.

A quarter revolution or 90 degrees later from the position shown in Fig. 110, the *plane* of the coil is perpendicular to the field and the two long conductors move *parallel to the lines of force*. Since the coil sides are not cutting any lines of force in this position, the *induced voltage is zero*. As the armature completes one-half of a revolution, the sides of the coil move once again *perpendicular* to the field and the induced voltage and current are again at a *maximum*. Since the left and right conductors of the coil are now interchanged, however, the direction of current flow is *reversed*, as you can verify with the generator rule. The current thus leaves at the brush marked (+) and returns through the brush marked (—) after flowing through the external circuit. Equivalently, since current flow is now from + to —, you might say that a *negative* current is flowing through the external circuit.

At the end of the *third* quarter of rotation, the long sides of the coil move again parallel to the lines of force and the induced voltage drops to zero. Finally, after completing a *full* revolution, the coil returns to its original position shown in Fig. 110, and the induced voltage and current are again at maximum values. Note that we have neglected the two short sides of the armature coil during the entire discussion, since they are not cutting across the magnetic field in any position.

Production of a Sine Wave. Let us look at the rise and fall of the induced voltage during one revolution of the armature coil a little more closely, since it is the basis of alternating-current generation. You will recall from our earlier discussion that a wire moving at some angle θ through a uniform magnetic field generates an emf equal to B L v sin θ × 10⁻⁸ volts. Referring to Fig. 111, let us redefine θ as the angle the plane of the generator coil forms with its *zero-emf* (vertical) position. Let us assign the symbol E_{max} to the *maximum* value of the emf, when the coil is horizontal. Since this maximum value is equal to B L v × 10⁻⁸, the induced emf, E, for any position of the coil is given by

$$E = B \, L \, v \times 10^{-8} \sin \theta = E_{max} \sin \theta$$

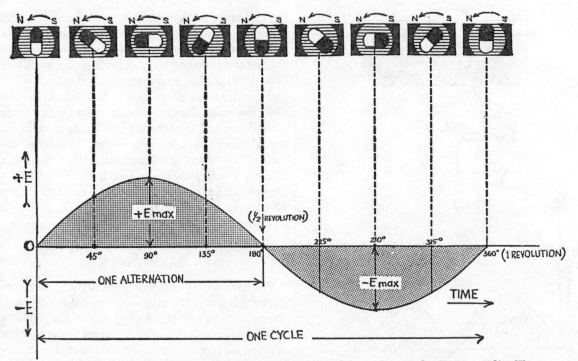

Fig. 111. A Coil Rotating in a Magnetic Field Generates an A-C Voltage or Current that Varies as a Sine Wave

In Fig. 111 we have plotted the equation $E = E_{max} \sin \theta$ for values of θ from 0 to 360 degrees. Just above this *sine wave* we have indicated the position of the armature coil (in cross section) corresponding to 45 degree increments in the angle of rotation. Since one complete revolution consists of 360 degrees, the schematic drawings show the coil positions at fixed instants of time, advancing counterclockwise in steps of one-eighth revolution (or 45°). The sine wave, therefore, also represents a graph of the **instantaneous values of the induced emf (E)** against time.

Fig. 111, which represents the induced voltage, as well as the induced current (in a closed circuit), shows that the current or voltage is *positive* for one-half revolution of the armature coil and *negative* for the other half-revolution of the coil. Such a (positive or negative) half-revolution is called an alternation and two such alternations make up a complete *cycle* (360°). Each cycle, corresponding to 360 degrees or a complete revolution of the coil, is of course exactly the same, oscillating between positive and negative peak values of E_{max} (called amplitude). The number of complete *cycles per second*, called the **frequency** of the sine wave, is the same as the number of revolutions per second of the coil (rps). The graph of Fig. 111 shows clearly that the **continuous rotation of a coil in a magnetic**

field naturally results in an induced alternating current or voltage that follows a sine wave. We will have much more to say about A.C. (Alternating Current) in the remaining chapters.

Direct-Current Generator. Many electrical applications, such as electroplating or charging batteries, require a source of **direct current,** flowing in one direction only. It is a simple matter to convert the basic a-c generator into a direct-current machine. All you need to do is to replace the two slip rings by the segments of a split ring or commutator of the type we discussed in connection with the direct-current motor (see Fig. 98). The automatic switching action of the commutator has the effect of *reversing* the *negative* half-cycles (alternations) of the a-c sine wave at the output of the generator, so that only **unidirectional** (positive) half-cycles remain. Fig. 112 illustrates the basic arrangement of the direct-current generator and also the unidirectional output waveform produced by it.

Note that the d-c generator of Fig. 112a is exactly the same as the motor shown in Fig. 98, except that the induced current is picked off the commutator *by* the brushes and fed to a load in an external circuit, rather than feeding current from an outside source *to* the commutator and armature coil. With the coil initially in the vertical position, as shown in Fig. 112a, the long sides are *parallel*

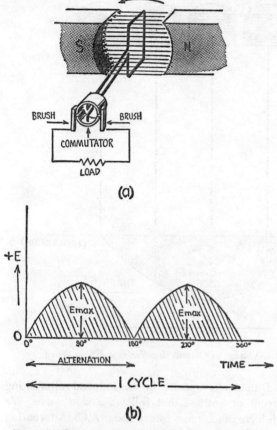

(a)

(b)

Fig. 112. Elementary Direct-Current Generator (a) and
Output Voltage Waveform (b)

to the flux, and since no lines of force are cut, the induced voltage is zero. Thus, we mark off the output voltage as zero at the start of the time interval on the graph of the output voltage waveform (Fig. 112b). With the coil turning counterclockwise, as indicated, it will be in a horizontal position one-quarter revolution later, and with the long sides cutting the flux perpendicularly, the induced (output) voltage rises to a maximum, as shown in Fig. 112b. After completing half a revolution, the sides of the coil are again parallel to the flux and the generator output voltage drops to zero, as shown in the graph. The output voltage of the d-c generator for the first half-revolution (one alternation) is exactly the same as that of the a-c generator, shown in Fig. 111.

Without the commutator, continued turning of the armature coil would interchange the positions of the two flux-cutting (long) sides, resulting in a reversal of the output voltage and current. At this very instant, however, the split segments of the commutator **interchange the output connections of**

the coil ends, so that the voltage induced (and the resulting current) is again in the *same direction*. As the coil completes its second half-revolution, the voltage (and current) rises and falls again exactly as before and in the *same direction* as during the first half-cycle. At the completion of the cycle (1 revolution) the commutator segments once again switch around the coil connections and another **unidirectional sine-wave half-cycle** results. We term this voltage waveform *unidirectional* rather than d.c. since it continually rises and falls. To obtain a *smooth* direct current, *filtering* is necessary. (See **Electronics Made Simple**.)

SELF INDUCTANCE AND MUTUAL INDUCTANCE

We have seen that a voltage is induced in a conductor whenever the magnetic flux linking the conductor is *changing*. If the magnetic field is due to a current-carrying conductor, a *change* in the number of lines of force of the field will accompany any change in the current. If the current through the conductor is increasing, the magnetic field is said to be *expanding;* if it is decreasing, the field is said to be collapsing. An *expanding* magnetic field may be caused by the closing of a switch that permits current to flow through a coil, while the opening of the switch would interrupt the current and cause the field to collapse (See Fig. 101). Expanding or collapsing fields may also be due to continuous variations of the current strength, such as may be brought about by *manually* rotating a rheostat in a d-c coil circuit or by connecting a coil to a source of *alternating current*. Regardless of the manner produced, whenever the lines of force of an expanding or collapsing magnetic field cut across a conductor (or the turns of a coil), an emf is induced.

You will recall, too, that by Lenz's Law the induced voltage is always of such a polarity as to *oppose the change of current that produces it.* Thus, when an applied voltage causes current flow in a coil circuit, the voltage induced in the coil will oppose the (change in) current and the applied voltage that caused it. For this reason, the induced voltage is also referred to as the **back emf or counter emf.** The characteristic property of a circuit that accounts for the production of an induced voltage or counter emf is called **inductance.** The greater the inductance in a circuit, the greater is its opposition to current changes and hence the greater the induced or counter emf. The schematic circuit symbol for inductance (*L*) is a coil (⟨coil symbol⟩), signifying that the property is primarily associated with coils.

Self-Inductance. Let us consider first a single coil of wire that is suddenly connected across an applied d-c voltage (Fig. 113). As the current rises in strength, an expanding magnetic field is established and an increasing number of lines of force cut across the turns of the coil. This in turn induces a counter emf in the coil that opposes the increase in the current and causes it to rise more *slowly* than it would without the magnetic field. As the field stabilizes the number of lines of force becomes constant, the counter emf drops to zero, and the current rises to its maximum value, determined by the applied voltage (*E*) and the resistance (*R*) of the coil (Fig. 113*b*). The distorted shape of the current and the time it takes to rise to its full value (E/R) is caused by the **self-inductance** of the coil. If the switch is now opened, disconnecting the battery and providing a short-circuit path for the current to flow

(Fig. 114*a*), the magnetic field will collapse and in the process again induce a counter emf that opposes the *decline* of the current. The short-circuit current of the coil therefore does not immediately drop to zero, but decays in an *exponential* manner, as shown in Fig. 114*b*.

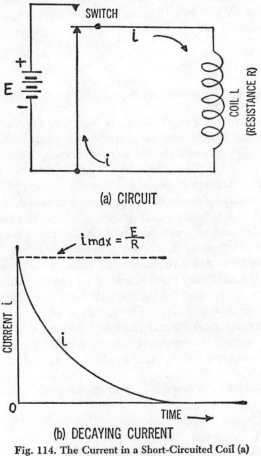

(a) CIRCUIT

$$i_{max} = \frac{E}{R}$$

(b) DECAYING CURRENT

Fig. 114. The Current in a Short-Circuited Coil (a) Decreases Exponentially to Zero (b)

If you connect the coil to an **a-c voltage source**, the alternating current will rise to a maximum, drop to zero, then rise again in the opposite direction and once more decline to zero, in accordance with the sine wave portrayed in Fig. 111. As a consequence of the continuous current variations, the magnetic field about the coil first builds up in one direction, then collapses to zero, builds up again in the *opposite* direction, and collapses once more, all in rapid sequence. This results in the **continuous induction** of counter electromotive forces that oppose the *variations* in the current flow and thus cause it to *lag behind* the applied voltage changes, as we shall later see in greater detail.

Magnitude of Induced (Counter) Emf. We have

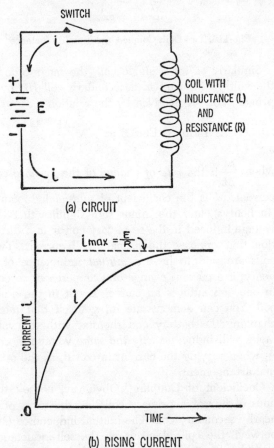

(a) CIRCUIT

$$i_{max} = \frac{E}{R}$$

(b) RISING CURRENT

Fig. 113. The Inductance of a Coil Suddenly Connected to a D-C Voltage (a) Causes the Current to Increase Relatively Slowly to its Maximum Value (b)

stated earlier that the emf induced in a conductor is proportional to the rate of change of flux linking the conductor. In a current-carrying coil, the **rate of change of flux is proportional to the rate of change of current** in the coil. We may now state more simply that the **counter emf induced in a coil is proportional to the rate of change of current through the coil**. Expressed as an equation, the (counter) emf

$$E = -L\frac{\Delta i}{\Delta t}$$

where $\frac{\Delta i}{\Delta t}$ represents the ratio of a small change in current to a small change of time (i.e., the rate of current change) and the proportionality constant, L, is called the **coefficient of self-inductance**, or simply **inductance**. (The minus sign signifies that the induced voltage opposes the applied voltage.) The formula also serves to define the **unit of inductance**, called the **henry** (after JOSEPH HENRY). The self-inductance of a coil (or circuit) is 1 henry if a current change of 1 ampere per second induces a counter emf of 1 volt in the coil. Smaller units, such as the millihenry (mh), representing one-thousandth of a henry, and the microhenry (μh), representing one-millionth henry, are frequently used.

EXAMPLE: A current change of 200 ma during a 0.1 second time interval induces a counter emf of 20 volts in a coil. What is its self-inductance?

Solution: $E = -L\frac{\Delta i}{\Delta t}$. Disregarding the minus sign,

$$20 = L\frac{0.2}{0.1} = 2\,L; \text{ hence}$$

$$L = \frac{20}{2} = 10 \text{ henrys.}$$

Mutual Inductance. We have already seen (Fig. 101) that the application of a voltage to a primary coil induces a momentary voltage in a secondary coil placed near it. Consider now two coils (A and B) placed close to each other, as shown in Fig. 115. The primary coil (A) is connected to battery through a rheostat to permit varying the coil current, while the secondary coil is connected to a voltmeter to indicate the induced voltage. When the current in the primary circuit is first established, the lines of force of the expanding field about coil A will link the turns of coil B and induce a *momentary* voltage in that coil, as shown by a "kick" of the

voltmeter. As soon as the primary current and field stabilize at their steady values, this voltage will disappear. If we now vary the current in the primary coil by moving the rheostat slider back and forth, a *variable flux* will thread the turns of the secondary coil and *induce a voltage* in it *proportional to the rate of change of the primary coil current*. The voltmeter will show the magnitude of the induced voltage. Coils A and B are said to be **coupled by mutual inductance**, or simply **inductively coupled**. If the primary coil is connected to an *a-c voltage*, the continuously varying primary coil current will, of course, induce a varying or a-c voltage in the secondary coil. More about that later.

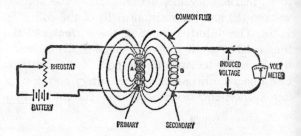

Fig. 115. Two Coils Coupled by Mutual Inductance

Similarly as for a single coil, the magnitude of the voltage induced in a secondary coil (B) by a primary coil (A) is given by the relation

$$\text{Induced Emf } E = -M\frac{\Delta i}{\Delta t}$$

where $\frac{\Delta i}{\Delta t}$ is the *rate of change* of the *primary* coil current, M is the coefficient of *mutual* inductance (in henrys), and the minus sign signifies that the voltage induced in the secondary coil is in a direction that *opposes* the primary coil current. Two coils are said to have a *mutual* inductance of 1 henry, if a current *change* of 1 ampere per second in one coil induces an emf of 1 volt in the other coil. You can demonstrate for yourself that interchanging the battery and rheostat with the voltmeter will induce *exactly the same* voltage in coil A, when varying the current in coil B, as the original arrangement.

Coefficient of Coupling. Obviously, not *all* the lines of flux of a primary coil link the turns of a nearby secondary coil. The mutual inductance (M) between the coils depends on the self-inductance of each coil and how *closely coupled* the two coils are. The mutual inductance may be made high by winding each coil with many turns, coupling the coils closely by bringing them physically close together, and by arranging their axes parallel to each other. The relation between the mutual inductance (*M*) between two coils of self-inductance L_1 and L_2, respectively, and the *coefficient of coupling*, *k*, is given by

$$M = k\sqrt{L_1\,L_2}$$

The coefficient of coupling, *k*, represents the relative amount of **flux interlinkage** between the coils, equivalently, or the absence of **flux leakage**. If all the flux produced by one coil links all the turns of the other coil, the flux leakage is zero and k = 1. This is the *tightest* possible coupling. If none of the flux of one coil links the other, k = 0, and there is no mutual inductance. (This may be achieved by placing the coils far away from each other and by placing their axes mutually perpendicular.)

EXAMPLE: Two coils of 4 and 16 henrys inductance, respectively, are tightly coupled with the coefficient of coupling k = 0.8. What is their mutual inductance?

Solution: $M = k\sqrt{L_1\,L_2} = 0.8\sqrt{4 \times 16} = 0.8 \times 8 = 6.4$ henrys.

The Induction Coil. Fig. 116 illustrates an induction coil, also often referred to as **spark coil** because of its use in producing sparks in automobile ignition systems. As illustrated, the device consists of a primary coil of relatively few turns of heavy wire wound around an iron core, and a secondary coil of many turns of fine insulated wire, wound in layers on top of the primary. The primary is connected in series with a battery, a switch (or key) and an **interrupting device**, similar to that described for the bell.

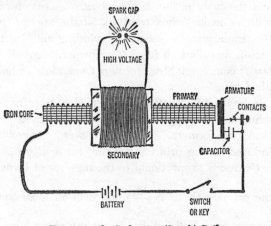

Fig. 116. The Induction (Spark) Coil

When the switch is closed, a direct current flows through the primary winding and magnetizes the core. While the current is rising, an expanding magnetic field is established and a voltage is induced in the secondary coil. An instant later, after the core has been sufficiently magnetized, the iron armature is attracted to the core and the primary circuit is interrupted at the contact points. To prevent the formation of an arc between the contact points due to the sudden interruption of the current (and high self-inductance of the primary), a small capacitor is placed in parallel with the contact points. The capacitor diverts a part of the arc current and thus assists in rapidly interrupting the circuit. The resulting sudden collapse of the magnetic field about the primary induces a very high voltage (of opposite polarity) in the secondary, which is aided by the many turns of fine wire linking the primary flux. (As we shall learn in connection with the transformer, the secondary voltage is proportional to the secondary-to-primary turns ratio.) As the primary core becomes demagnetized, the armature returns to its original position and closes the primary circuit again, thus repeating the entire cycle. The rapid buzzer-type action results in many interruptions of the primary current each second, causing the induction of a very high secondary voltage, constantly changing in polarity. This is a type of A.C.

An **automobile induction coil** typically consists of several hundred primary turns and up to 20,000 secondary turns. It is capable of boosting the 6 or 12-volt battery potential to about 20,000 volts, which are applied to the spark-gap terminals of the spark plug, where they ignite the gasoline mixture in the familiar process. The spark coil must furnish about 200 sparks per second in a car traveling at 60 miles per hour.

Transformers. Invented in 1886 by WILLIAM STANLEY, the **transformer** is the most important induction device. Transformers are capable of *stepping up* an a-c voltage to very high values, permitting the transmission of large amounts of power over long cables without undue voltage (IR) losses. By stepping up the voltage at the generator to values close to a half million volts, the current sent over the power line can be relatively small for a given amount of required power (P = E × I), permitting a reduction in the size of the cables. At the receiving end of the power line, the voltage is then reduced by another (*step-down*) transformer to a value suitable for homes, offices, and factories.

A transformer consists essentially of two coils coupled by mutual inductance (see Fig. 117). The

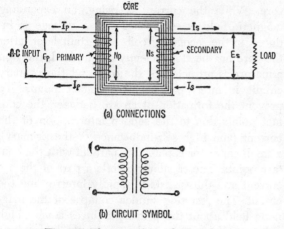

(a) CONNECTIONS

(b) CIRCUIT SYMBOL

Fig. 117. Elements of Simple Transformer

coils are electrically insulated from each other, but are linked by common magnetic flux. One coil, the **primary winding,** is connected to the a-c voltage supply (generator), while the other coil, called the **secondary winding,** is connected to a **load,** which may be any electrical device whatever. The transformer thus transfers electrical energy from the primary circuit to the secondary circuit without a direct connection and permits at the same time a step-up or step-down of the primary voltage or current. The magnetic flux in a transformer may link the coils either through an *iron core* or an *air core,* the latter being used at relatively high a-c frequencies (called **radio frequencies**). Iron core transformers are generally either of the **core type,** with the coils encircling the iron core; or of the **shell type,** with the core surrounding the coils (Fig. 118).

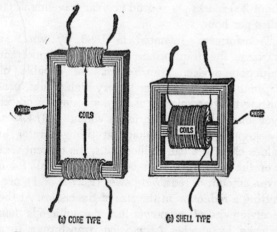

Fig. 118. Construction of Core-Type (a) and Shell-Type (b) Iron Core Transformers

Ideal Transformer. With the primary winding connected to an a-c supply, the alternations of the primary current set up an *alternating* magnetic field in the core that is continually expanding, collapsing, and building up again in the opposite direction. This alternating flux induces an alternating (a-c) voltage in the secondary winding, and can supply a current through a closed secondary circuit. The variations in the flux, which produce the secondary emf, also affect the primary winding (due to its self-inductance), and by Lenz's Law, induce in it a counter emf that opposes the a-c voltage applied to the primary winding. The value of this counter emf is *almost equal* to the applied voltage, when no current is drawn from the secondary winding and, hence, very little current flows through the primary under **no-load** conditions. The small current that does flow is known as the no-load or **magnetizing current,** since it magnetizes the core. When a current is drawn by the secondary load, a proportional current flows through the primary, as we shall see later. If the primary of a transformer is connected to a **d-c voltage,** a voltage is induced in the secondary for the *instant* during which the magnetic field is building up, but this voltage collapses immediately, as soon as the field reaches a steady (unchanging) value. Because of the absence of a counter emf for d.c., the primary current will be large, and since the d-c resistance of the winding is small, the primary winding will **burn out.** The transformer, thus, is strictly an a-c device; **never connect it to d.c.!**

In a transformer having a closed iron core practically all the lines of force produced by the primary winding link every turn of the secondary winding and the leakage flux is almost zero. A transformer without leakage flux transfers all the energy from the primary to the secondary winding and, for this reason, is called an **ideal transformer.** Some of the larger commercial transformers come close to being ideal transformers.

A few simple relations hold for ideal transformers which are also approximately correct for most practical transformers. As we have seen, the voltage induced in the primary winding for a *given* magnetic flux is **proportional to the number of turns** of the primary ($E = N \dfrac{\Delta\phi}{\Delta t}$). Since in an ideal transformer (one without flux leakage) every turn of the secondary is linked by this same magnetic flux, the voltage induced in the secondary winding is proportional to the number of turns in the secondary. It follows that for an ideal transformer the ratio of the primary to the secondary voltage equals the ratio of the number of turns in the two windings. Stated conveniently in mathematical form:

$$\frac{E_p}{E_s} = \frac{N_p}{N_s}$$

where E_p and E_s are the primary and secondary voltages, respectively, and N_p and N_s are the number of primary and secondary turns, respectively. This formula obviously does *not* apply to an air-core transformer, where considerable flux leakage exists.

EXAMPLE: An (ideal) iron-core transformer has a primary winding of 500 turns and a secondary winding of 3500 turns. If 115 volts a-c are applied to the primary, what is the voltage across the secondary?

Solution: $E_s = \dfrac{N_s}{N_p} \times E_p = \dfrac{3500}{500} \times 115 = 7 \times$ 115 = 805 volts.

Equivalently, you might consider that the transformer has a *step-up* ratio of 500:3500 or 1:7; hence, the secondary voltage is seven times as great as the primary, or $E_s = 7 \times 115 = 805$ volts.

If no energy is lost by leakage flux (and other causes) the power output of an ideal transformer must be the *same* as the power input to the primary winding. Hence, we can write (since $P = E \times I$)

$$E_p I_p = E_s I_s$$

or

$$\frac{I_p}{I_s} = \frac{E_s}{E_p}$$

but from the previous relation:

$$\frac{E_s}{E_p} = \frac{N_s}{N_p}$$

Substituting:

$$\frac{I_p}{I_s} = \frac{N_s}{N_p}$$

Stated in words, the **primary-to-secondary current ratio** is equal to the reciprocal of the primary-to-secondary turns ratio. By comparison with the previous formula, it is apparent that the **current is stepped down**, whenever the voltage is stepped up, and vice versa.

EXAMPLE: If the primary current in the previous example is 3.5 amperes, what is the current flowing in the secondary winding?

Solution: $I_s = \dfrac{N_p}{N_s} \times I_p = \dfrac{500}{3500} \times 3.5 = \dfrac{3.5}{7} =$ 0.5 ampere.

Equivalently, since the voltage step-up is 1:7, the current step-*down* is 7:1, or $3.5/7 = 0.5$ ampere.

Practical Transformers. In contrast to ideal transformers, **real transformers** depart somewhat from the relations we have discovered for ideal transformers. It is actually not possible to build a transformer with *unity* coupling coefficient, as we have assumed above, and some leakage flux is always present. Consequently, the voltage (and current) transformation ratio is always a little *less* than the turns ratio. In an iron-core transformer, this discrepancy usually can be neglected with very little error. In an air-core transformer, however, the flux leakage is considerable (k is much less than 1), and the simple voltage and current ratios for ideal transformers do not hold.

Actual transformers, moreover, have a certain voltage (IR) drop in the **resistance** of their windings. Hence, the voltage across the primary coil is a little less than the applied a-c voltage and the induced voltage in the secondary is a little greater than the voltage actually available at the load terminals. A certain amount of power is thus lost in the resistance of the primary and secondary windings. This power (I^2R) loss, called the **copper loss**, is wasted in *heating* the windings.

In addition to the copper loss, there is an **iron loss** within the iron core of the transformer. The total iron loss is caused by two factors: **hysteresis losses** and **eddy current losses**. Hysteresis, you will recall, is the lagging of the magnetic flux produced in the core behind the magnetizing force. Energy is required to align the domains in the magnetic core material, and this energy must be supplied by the power source, thus constituting a power loss. The smaller the area of the hysteresis loop of the magnetic material used for the core, the smaller are the hysteresis losses.

Eddy currents are small circulating currents set up in a large conductor, either by motion of the conductor in a magnetic field or by a varying magnetic field in a stationary conductor. The changing magnetic field in a transformer induces an emf not only in the windings, but also *in the core itself*. The emf induced in the core, in turn, gives rise to extraneous circulating (eddy) currents, which heat up the core and result in a power loss. Eddy currents are proportional to the **square of the core**

thickness, which gives the clue to almost eliminating them. By dividing the core into a bundle of thin magnetic sheets, called **laminations**, eddy current losses may be made negligible. Some transformers use suspensions of many small iron particles, known as **powdered iron,** to achieve the same purpose. Unfortunately, eddy current losses are also proportional to the square of the frequency. It becomes impossible, therefore, to use ordinary iron cores for transformers used at radio frequencies, since the losses would be excessive. Radio-frequency transformers, for this reason, usually have air cores or sometimes very finely powdered iron cores.

Eddy currents set up in large conductors moving through a magnetic field are used to advantage as "magnetic brakes" in certain instruments, such as the watt-hour meter. By Lenz's Law, the currents set up in the conductor oppose its motion through the field and thus provide a braking effect.

The sum total of the copper and iron losses in a transformer prevent its power output from equaling the power supplied. The actual performance of a transformer is measured by its **efficiency,** which is defined as the **ratio of power output** (from the secondary) **to power input** (to the primary). Actual transformers realize efficiencies of 90 to 98 percent, the higher figure applying to large commercial units. Because of the losses, the current ratio of an actual transformer is *not* equal to the reciprocal of the turns ratio, but must be *multiplied* by the efficiency of the transformer. The voltage ratio may be assumed equal to the turns ratio, since the copper losses usually may be neglected.

EXAMPLE: A 90-percent efficient transformer has a primary winding of 1000 turns, a secondary winding of 200 turns, and an a-c voltage of 120 volts applied to the primary. If the primary supplies a current of 2 amperes, what current will be available from the secondary winding? What is the secondary voltage?

Solution: For an *ideal* transformer, the secondary current would be $I_s = \dfrac{N_p}{N_s} \times I_p = \dfrac{1000}{200} \times 2 = 10$ amps; For an efficiency of 90 percent, the current *actually* delivered by the secondary will be $0.9 \times 10 = 9$ amperes. The secondary voltage $E_s = 200/1000 \times 120 = 120/5 = 24$ volts.

The Autotransformer. In addition to the many applications of transformers in power distribution, radio and electronics, one unconventional type of transformer combines the primary and secondary into a **single tapped** winding (see Fig. 119). The arrangement is called an **autotransformer.** Either step-up or step-down voltage ratios may be obtained. The type illustrated in Fig. 119 is a *step-down* transformer, since the input voltage is applied across the entire winding, serving as primary, while the output voltage is taken from the portion of the winding included between one end and the tap. The autotransformer does *not* provide isolation between primary and secondary circuits, but its simplicity makes it economical and space-saving. In one type of autotransformer, known as **Powerstat** or **Variac,** the winding is arranged into circular form and the tap is made adjustable by rotating a sliding contact along the winding. Almost continuous control of the step-down or step-up ratio, and hence of the output voltage, can be achieved in this way.

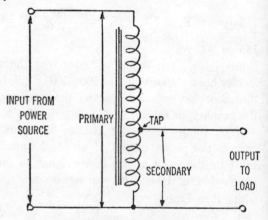

Fig. 119. Schematic Diagram of Step-Down Autotransformer

Practice Exercise No. 10

1. State the three ways that Faraday found for increasing the magnitude of the induced current (and emf) when moving a magnet toward a coil connected to a galvanometer. (See Fig. 100.)

2. What did the *direction* of the induced emf depend upon in Faraday's experiments?

3. State the two laws of induction based on Faraday's generalizations. Are they equivalent to each other?

4. A coil wound with 20 turns of wire is placed horizontally on a table and the north pole of a magnet is moved toward it. If the magnetic flux linking the coil changes from 2000 lines to 100,000 lines in 0.1 second, what is the magnitude and direction of the average emf induced in the coil?

5. An emf of 6 volts is induced in a 200-cm long wire that is moved at a speed of 12 meters per second at right angles across a uniform magnetic field. What is the flux density?

6. A 30-cm long rod travels at a speed of 10 meters per second across a magnetic field of 20,000 gauss. If the rod makes an angle of 60° with the vertical perpendicular to the lines of flux (30° with the horizontal), what voltage will be induced between its ends?

7. State Lenz's Law. On what law of nature is it based?

8. Can you predict the direction of the emf induced in the generator coil in Fig. 110 and the direction of current flow through the load? Using both Lenz's Law and the left-hand generator rule, do you obtain the same result?

9. State the essential difference between a generator and a motor; between a d-c and an a-c generator.

10. Describe the evolution of a sine wave voltage and current in an a-c generator. What happens to the output waveform, when the slip rings are replaced by a commutator?

11. What is *inductance?* Distinguish between *self-inductance* and *mutual* inductance.

12. Why is there a spark when the circuit of a current-carrying coil is suddenly opened?

13. Why does the current lag behind the voltage when a coil is *first* connected to d.c., or when it is connected to a.c.?

14. If a voltage of 30 volts is induced in one of two mutually coupled coils, when the current in the other is changing at the rate of 5 amperes per second, what is the mutual inductance?

15. Two coupled coils of 3 and 12 henrys self-inductance, respectively, have a mutual inductance of 4 henrys. What is the coefficient of coupling (*k*)?

16. What is an *induction coil?* State the essential difference between an induction coil and a transformer. Which device is more efficient, in your opinion?

17. Describe the operation of a transformer and develop the voltage and current relations for an ideal transformer.

18. Two thin and two heavy wires protrude from the terminals of an unlabeled transformer. Which pair of wires is designed to carry the high voltage?

19. A transformer steps down the 120-volt a-c supply voltage to 6 volts for the filaments of radio tubes. What is the primary-to-secondary turns ratio? If the primary has 1000 turns, how many turns are wound on the secondary?

20. State the losses incurred in actual transformers. Distinguish between *copper* and *iron* losses.

21. What are *eddy currents?* How can they be reduced or eliminated?

22. Define the *efficiency* of a transformer. What efficiency values are likely to be attained in practice?

23. If 120 volts are applied to the primary of an 85% efficient transformer, the secondary voltage is 600 volts and the secondary current is 0.17 ampere. What is the primary current?

SUMMARY

Laws of Induction: 1. An electromotive force is induced in a coil of wire whenever the magnetic flux linking the coil is *changing;* the *magnitude* of the induced emf is proportional to the *rate* at which the number of lines of force through the coil are *changing*.

2. An emf is induced in any conductor that is *cutting across lines of force;* the magnitude of the induced emf is *proportional to the rate* at which lines of force are being cut.

An emf of *1 volt* is induced in a conductor or single loop, when it cuts magnetic lines of force at the rate of 100,000,000 per second. For a coil of *N* turns this emf is induced in each turn, and the total emf is the rate of change of flux (in webers) times the number of turns. ($E = N \dfrac{\Delta\phi}{\Delta t}$ volts). If the flux is given in gauss, the total emf $E = N \dfrac{\Delta\phi}{\Delta t} \times 10^{-8}$ volts.

When a conductor of length *L* moves across a magnetic field of flux density *B* at a velocity *v* and making an angle θ with the horizontal, the emf induced in the conductor is given by

$$E = BvL \sin \theta \times 10^{-8} \text{ volts}$$

Lenz's Law: When a current is set up by an induced emf due to the motion of a closed-circuit conductor, the *direction* of the current will be such that its magnetic field will *oppose* the motion.

Left-Hand Generator Rule:
Fore (Index) Finger = *Flux*
Thumb = *Motion*
Center (Middle) Finger = Direction of *C*urrent or *E*mf.

An **electric generator,** consisting of an armature coil rotating in a magnetic field and means for connecting the coil to an external circuit, **converts mechanical work into electric energy.** An a-c generator is provided with continuous **slip rings,** on which the brushes ride, while a d-c generator has a split-segment commutator. An a-c generator naturally produces a **sine-wave** current or voltage, while a d-c generator produces a **unidirectional** emf or current (half-sinewave).

Self-inductance accounts for the production of a counter emf in a coil that *opposes* the applied voltage. The greater the inductance, the greater the counter emf. Inductance opposes any *change* in the current, causing it to *lag behind* the voltage.

The counter emf induced in a coil is *proportional* to its *inductance* and to the *rate of change of the current* through the coil ($E = -L\dfrac{\Delta i}{\Delta t}$). The self-inductance of a coil is *1 henry* if a current *change* of *1 ampere per second* induces a counter emf of *1 volt* in the coil.

Mutual inductance exists between any two **coupled** coils. The mutual inductance between two coils is 1 henry, if a current *change* of 1 ampere/second in one coil induces an emf of 1 volt in the other coil. Magnitude of induced emf $E = -M\dfrac{\Delta i}{\Delta t}$.

The **coefficient of coupling** between two coils (k) indicates the relative **tightness** of coupling and determines the mutual inductance. ($M = k\sqrt{L_1 L_2}$).

A **transformer** consists of a primary and a secondary coil linked by common magnetic flux through an iron-core or air core. The **primary** is connected to an **a-c voltage**, the **secondary** to a **load**.

In an **ideal** transformer, the **primary-to-secondary voltage ratio equals the** (primary-to-secondary) **turns ratio**; and the primary-to-secondary **current ratio** is equal to the **reciprocal** of the **turns ratio**.

$$\left(\frac{E_p}{E_s} = \frac{I_s}{I_p} = \frac{N_p}{N_s} \right)$$

A **practical** transformer has losses due to the resistance of its windings (**copper loss**), due to hysteresis and eddy currents (**iron loss**); as a result its efficiency (output/input) is less than 100 percent and its output current is reduced by the efficiency.

$$(I_s = I_p \times \frac{N_p}{N_s} \times \text{efficiency}).$$

An **autotransformer** combines the primary and secondary windings into a single, tapped winding, at the expense of circuit isolation.

CHAPTER ELEVEN

ALTERNATING CURRENT FUNDAMENTALS

We have seen how the continuous rotation of the armature coil in the magnetic field of the generator naturally generates an **alternating voltage** or **current** (in closed circuit) that rises and falls in magnitude as a sine wave. We shall now consider the fundamental relations and definitions that hold for sine-wave currents and voltages. Then, we shall be ready to deal with **alternating-current circuits**—containing inductance, capacitance, and resistance—almost as readily as we solved d-c circuits by means of Ohm's Law.

A-C DEFINITIONS

Remember that only the **projection** of the armature coil that cuts the flux *vertically* is responsible for inducing an emf in the coil, and that this vertical projection varies as the *sine* of the angle of rotation (sin θ). Let us now reconstruct the a-c sine wave once more by plotting the **vertical projections** of the long (flux cutting) sides of the armature coil against time or the number of degrees of one revolution. To obtain a simplified picture, assume that the sides of the armature coil are represented by the *length of the radius* in the circle shown in Fig. 120 and that this length represents the maximum value (E_m) of the alternating voltage. Let

the radius turn counterclockwise in the circle to simulate the actual rotation of the armature in an a-c generator. The instantaneous emf (symbol e) induced for *any* position of the armature coil is then represented by the vertical projection of the rotating radius, which is a line drawn from the end of the radius *perpendicular* to the horizontal diameter of the circle. The length of this perpendicular line (projection) at any time is, of course, equal to the *sine* of the angle the radius forms with the horizontal diameter (see Fig. 120).

To the right of the rotating radius in Fig. 120 we have plotted the length of the vertical projection against the counterclockwise angle of the radius with the horizontal diameter. Thus, when the rotating radius makes an angle of 30° with the horizontal (point 2), a horizontal line drawn from the intersection of the radius with its vertical projection to the 30°-ordinate of the waveform at right determines the height of the projection for 30° rotation and, hence, the value of the voltage at that instant. Similarly, when the radius makes an angle of 90° (point 4), the horizontal line drawn to the 90° ordinate of the voltage waveform determines the maximum value of the voltage (E_m) at that instant, for 90° rotation. Evidently, as the radius rotates, its vertical projection varies between maxi-

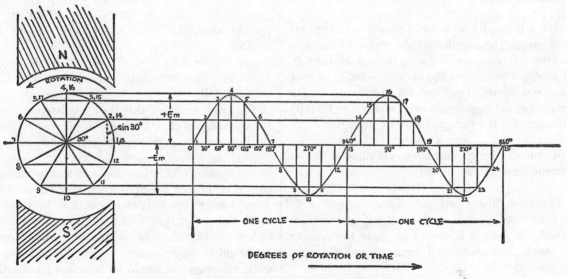

Fig. 120. Generation of Sine Wave Voltage by Rotating Radius

119

mum values of $+E_m$ and $-E_m$ and generates the sine wave of voltage shown at right of Fig. 120. After one complete rotation, or *1 cycle*, the sine wave repeats itself during the next rotation, and so on. The waveform of the instantaneous current (*i*) in a closed circuit is, of course, exactly the same.

Cycle, Frequency and Period. We have seen that during the time the armature coil rotates through 360 degrees or one revolution, the output voltage goes through a complete **cycle**, consisting of a **positive alternation** (the first 180 degrees) and a **negative alternation** (the second 180 degrees). During each alternation the voltage attains a maximum value, also called the **amplitude** or **peak voltage**, which is positive ($+E_m$) during the positive alternation and negative ($-E_m$) during the negative (second) alternation of each cycle. The time required to complete one full cycle (two alternations) is called the **period**, and the number of cycles completed per second is called the **frequency** of the sine wave. Period (*T*) and frequency (*f*) are inversely related to each other; that is, frequency is the *reciprocal* of the period (f = 1/T) and vice versa. For example, if a coil rotates between two poles of an electromagnet at a speed of 3600 revolutions per minute, or 60 rev. per second, it will generate A.C. at a frequency of 60 cycles per second (cps) and the period will be T = 1/f = 1/60 second.

Angular Velocity. We have already developed the mathematical expression for the instantaneous value of the a-c sine wave voltage (e) in terms of the peak value, E_m, and the angle or rotation, θ, which is also known as the **angular displacement**:
$$e = E_m \sin \theta$$
where θ is usually expressed in *radians* rather than in degrees. The angular displacement or angle of rotation (θ) completed during a certain time (t) depends, of course, on the **angular velocity** of rotation (symbol w). Thus, we may replace θ by the product of the angular velocity (w) and the time (t), obtaining for the instantaneous a-c voltage
$$e = E_m \sin wt$$
and similarly, for the instantaneous value of the a-c current (in a closed circuit)
$$i = I_m \sin wt$$
Moreover, since each revolution comprises 2π radians (360 degrees = 2π radians), the angular velocity *in radians* is simply 2π times the number of revolutions completed each second (i.e., the frequency)
$$w = 2\pi f = 6.283 f$$

EXAMPLE 1: A coil rotates between the 2-pole field of an a-c generator at the rate of 3600 r.p.m. If the peak value of the induced emf is 220 volts and the peak value of the current through a load is 10 amperes, write the expressions for the instantaneous values of the voltage and current at any time.

Solution: A speed of 3600 r.p.m. is equivalent to 60 r.p.s.; the frequency, therefore, is 60 cycles per second. Hence, the angular velocity (in radians) $w = 2\pi f = 6.283 \times 60 = 377$.
Thus, $e = 220 \sin 377 t$
and $i = 10 \sin 377 t$

EXAMPLE 2: If the peak value of an a-c voltage is 100 volts and its frequency is 400 cps, what are the *instantaneous* values of the voltage 0.000625 sec, 0.00125 sec, 0.001875 sec and 0.0025 seconds after the generator is first turned on? (Assume that the generator starts with zero voltage at zero time.)

Solution: $e = E_m \sin wt = 100 \sin 2\pi \times 400 t =$ $100 \sin 800\pi t$
Hence, after 0.000625 sec: $e = 100 \sin 800\pi \times .000625 = 100 \sin 0.5\pi$ (rad) =

$$100 \sin \frac{\pi}{2} (rad) = 100 \sin 90° = \textit{100 volts.}$$

after 0.00125 sec: $e = 100 \sin 800\pi \times .00125$
$$= 100 \sin \pi \text{ (rad)} = 100 \sin 180°$$
$$= 100 \times 0 = \textit{0 volts.}$$
after 0.001875 sec: $e = 100 \sin 800\pi \times .001875$
$$= 100 \sin 1.5\pi \text{ (rad)}$$
$$= 100 \sin \frac{3\pi}{2} \text{ (rad)}$$
$$= 100 \sin 270°$$
$$= 100 \sin (-90°)$$
$$= 100 \times -1 = \textit{-100 volts}$$
Finally after 0.0025 sec (1/400 sec), or one full cycle:
$$e = 100 \sin 800\pi \times 0.0025 = 100 \sin 2\pi \text{ (rad)}$$
$$= 100 \sin 360° = 100 \sin 0° = 100 \times 0$$
$$= \textit{0 volt}$$
The example shows that the a-c voltage oscillates between values of 0, 100, and -100 volts.

Effective (Root Mean-Square) Value of A.C. Though an a-c sine wave makes a pretty picture, its continuous oscillations make it somewhat difficult to determine exactly a particular value of the voltage or current that can be said to be *effective* in an a-c circuit. Just which of the many possible instantaneous values of voltage or current would or should an electric meter read? The best way to settle this question would be to define an **effective value** of alternating current that would perform *work at exactly the same rate* as an equal value of direct current. This is precisely what has been done and the only question is, just what is this effective value of an a-c current or voltage?

The easiest way to compare the rates of work (power) of an alternating current and a direct current is to measure the relative **heating effect**, when each flows through the same value of resistance. Accordingly, we define the **effective value** of an alternating current as that a-c value which produces heat at exactly the same rate as an equal amount of direct current flowing through the same resistance. In other words, an *effective* value of 1 ampere *a.c.* will produce the same heat in a given resistor and given time as 1 ampere *d.c.*

With this definition it is easy to compute the effective value of an alternating current. We know that heat production is proportional to the *square* of the current for a given resistance (heating rate = power = I^2R). So, let us *square* all the instantaneous values (ordinates) of an alternating-current sine wave, as illustrated in Fig. 121. Here the top graph shows a typical a-c sine wave of the instantaneous current (i) against time, varying between peak values of $\pm I_m$. The bottom graph of Fig. 121, illustrates the *squared* sine wave obtained when all the instantaneous values of the current (i) in the upper graph are squared and the corresponding i^2 values are then plotted against time. Note that the lower graph, because of the squaring process, has only *positive* values that oscillate between zero and I_m^2 about a new axis. Since the curve varies uniformly between these extreme values (0 and I_m^2), its average or *mean value* must be equal to $\frac{1}{2}I_m^2$ (i.e., $\frac{0 + I_m^2}{2} = \frac{1}{2}I_m^2$). We now need only extract the *root* of this *mean-squared* value ($\frac{1}{2}I_m^2$) to obtain the effective a-c value in accordance with our definition. This value is frequently called the root-mean-square or rms value. Thus, the effective, or rms value (I) is

$$\sqrt{\frac{I_m^2}{2}} = \frac{I_m}{\sqrt{2}} = \frac{I_m}{1.414} = 0.707 \, I_m$$

Hence, for an a-c current:

effective (rms) value $I = 0.707 \, I_m$;

and, similarly, for an a-c voltage:

effective (rms) value $E = 0.707 \, E_m$.

If you want to determine the *maximum* values from given effective values of the voltage or current, simply take the *reciprocals* of these relations. Thus,

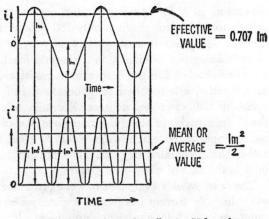

Fig. 121. Determining the Effective Value of an Alternating Current

$$E_m = 1.414 \, E$$
and $I_m = 1.414 \, I$

where E and I are the effective values of the voltage and current respectively. (Effective values are usually stated without a subscript.)

EXAMPLE: When an effective a-c voltage of 115 volts is applied to a circuit, a *peak* value of 28.3 amperes current is observed to flow. What is the peak value of the voltage and the effective (rms) value of the current?

Solution:

$$E_m = 1.414 \, E = 1.414 \times 115 = 162.8 \text{ volts}$$
and $I = 0.707 \, I_m = 0.707 \times 28.3 = 20$ amperes.

Average Value of A.C. When the value of an a-c voltage or current is stated without specific designation, the effective or rms value is always meant. Occasionally the *average* value of an alternating current or voltage is referred to; this does *not* mean the effective (rms) value. Looking at the sine wave illustrated in Fig. 121 (top), it is evident that the **average value of an alternating current over one complete cycle is zero**, since the curve oscillates uniformly about the X-axis or zero. The average value, therefore, is always taken over one-half of a cycle (one alternation), and it can be demonstrated mathematically that it is equal to 0.636 of the maximum value. Summarizing the relations between average (I_{av} or E_{av}), peak (I_m or E_m) and effective or rms values (I or E):

$$I_{av} = 0.636 \, I_m = 0.9 \, I \text{ (rms)}$$
and $E_{av} = 0.636 \, E_m = 0.9 \, E$ (rms)

Phase, Phase Angle and Phase Difference. Up to now we have assumed that the rotation of the armature coil through 360 *geometrical* degrees, or one revolution, will always generate *one* cycle (360°) of a-c voltage. Actually, this is only true for a *two-pole* generator. In a four-pole generator, for example, an armature rotation through only 180 geometrical degrees will generate a complete a-c cycle, or 360 *electrical* degrees. Similarly, in a six-pole machine only *one-third* or 120 geometrical degrees of armature rotation are required to generate a full a-c cycle, or 360 *electrical* degrees. For this reason, the degree markings along the sine-wave axis of an a-c voltage or current always refer to **electrical** rather than **geometric** degrees. They are the same only in the case of a two-pole generator.

The *fraction* of a cycle that has elapsed since an a-c voltage or current has passed a given reference point—measured in electrical degrees—is also referred to as the **phase** or **phase angle** of the voltage or current. The reference point, from which the phase angle is measured, usually is the starting point of the voltage or current waveform, or *zero* electrical degrees. Thus, at the start of the a-c voltage in Fig. 120 (point 1), the phase is *zero*. At point 2 the phase or phase angle is 30°, at point 3 it is 60°, at point 4 it is 90°, and so on throughout the complete cycle until point 13, where the phase angle is 360° or 0°. When used in this way, the term **phase** is significant only for a fraction of one cycle or one complete cycle. The phase repeats during each cycle.

The term **phase angle** or **phase difference** is more commonly used to compare *two* a-c voltages, currents, or a voltage and a current of the *same frequency* that pass through zero values at different instants. Since, with the frequency the same, each a-c cycle occupies exactly the same amount of time, the *phase difference* between two such voltages or currents (or a voltage and a current) is conveniently expressed in *fractions of a cycle* or electrical *degrees*, time being implied in either case. Thus, waveform 1 (solid line) in Fig. 122a represents some a-c voltage or current and waveform 2 (dotted line) represents another voltage or current of the *same* frequency that passes through zero at different instants than waveform 1 and, hence, is said to be **out of phase** with it. At the 0° reference point waveform 1 has a value of zero, while waveform 2 is at its negative maximum at this instant and does not reach zero until the 90° point on the axis. At this point, however, waveform 1 has already reached its positive maximum value. Clearly, waveform 1 is out of phase with waveform 2 by one-quarter cycle or 90°. Moreover, since waveform 1 reaches corresponding points of the cycle (such as maximum and minimum points) *earlier* than wave-

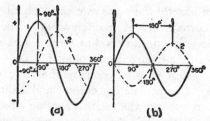

Fig. 122. Phase Differences Between Sine Waves

form 2, it is said to *lead* waveform 2 *in phase*, in this case, by 90°. Equivalently, waveform 2 is said to *lag* waveform 1 by 90° or one-quarter cycle.

Fig. 122b shows two sine waves of the same frequency that are 180° or one-half cycle out of phase. Here curve 1 rises in the positive direction from the 0° starting point, while curve 2 starts to go negative at the same point. Measuring the phase difference between the positive maximum points, the dotted curve (2) is seen to reach its maximum at the 270-degree marker, while the solid curve (1) reaches it at the 90-degree marker. The phase difference, therefore, is 270° − 90° or 180°. We could have measured the phase difference at any other two corresponding points of the cycle, such as the starting points, with the same result; it does not matter where the phase is measured. Also, in this particular case, where the sine waves are 180° (one-half cycle) out of phase, either wave may be considered leading or lagging. Curve 2 is always positive when curve 1 is negative, and vice versa.

USE OF VECTORS

The portion of the world around us that is controllable is *quantitative*. Quantities permit calculation and manipulations, and thus control. Not all quantities can be expressed as a single number or magnitude; some have *both magnitude and direction*. Quantities that have magnitude only are called **scalar** and those that have both magnitude and direction are called **vector** quantities. There is an abundance of both types of quantities, though you may not have been aware of it. Length, height and width, for example, are *scalar* quantities, since *one number* is sufficient to specify any one of them. *Force* and *velocity* in contrast, are *vector* quantities, since they have both magnitude *and* direction. If two people pull a load with the same *magnitude* of force, they may not be helping each other much, unless they pull in similar *directions*. If they pull in opposite directions the magnitude of their forces, obviously, will cancel out, as you know from rope pulling contests. The *speed* (a scalar) of a car may be 60 m.p.h., but to know its *velocity* (a vector) you must specify in which *direction* the car is going at 60 m.p.h. Given a starting point and the velocity—say 60 m.p.h. in a north-easterly direction—you can tell just where the car will be at any time.

Calculations in electricity involve both scalar and vector quantities. *Potential* or voltage, for example, is a *scalar* quantity. If the emf of a source is specified as 220 volts, you do not ask in which direction the voltage is going. In contrast, electrostatic or magnetic *forces* are *vector* quantities. If two charges exert a force of 2000 dynes upon each other, you will not know what is happening unless the *direction* of the force is specified. You must know whether the charges are repelling or attracting each other and whether they tend to move up, down, or sideways. You can think of many more examples for yourself.

The reason we have not mentioned the difference between scalars and vectors before is that we have been dealing primarily with scalar quantities, which can be added, subtracted, multiplied or divided just like ordinary numbers. Our present interest in vectors is due to the fact that **most a-c quantities involve both magnitude and a phase angle** (direction) and hence are best represented by *vector* quantities. For example, the 90-degree phase difference between the two sine waves in Fig. 122*a* can be represented by two *straight lines* making a right angle (90°) with each other, and the 180° phase difference in Fig. 122*b* by two lines going in opposite directions. This is a typical application of vectors to A.C. and it is much simpler than drawing the curves of Fig. 122. We shall have more to say about this later.

Vector Representation. A **vector** quantity is generally represented by a **straight line that is pointed in a specific direction.** The **length** of the line denotes the **magnitude** of the vector quantity and its **orientation** with respect to some base or reference line denotes the **direction** of the vector quantity. Fig. 123 illustrates some typical vector representations. The general appearance of a vector is shown in (*a*) of the figure. Here the length of the line segment *OA* represents the *magnitude* of some vector quantity, while its angular orientation and the arrow point (to the right) stands for the *direction* of the vector quantity. Fig. 123*b* shows the representation of *geographical location* by means of a vector. Here point *O* marks the position of an observer (on a map) and point *X* represents a certain loca-

tion, which he would like to reach. The length of the line (3 inches), on a scale of 1 inch per mile, represents the fact that the desired place is 3 miles away from the observer. The direction of the line (north-west on the page) indicates that the place is north-west from his present position.

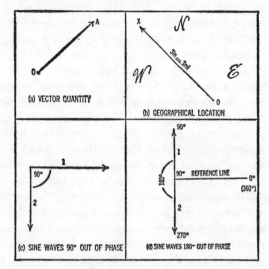

Fig. 123. Vector Representation: (a) Vector Quantity; (b) Geographical Location; (c) Sine waves 90° out of phase and (d) Sine Waves 180° out of phase

Fig. 123*c* shows the vector representation of the two 90-degree out-of-phase sine waves illustrated in Fig. 122*a*. Note that the greater length of vector *1* represents the greater *amplitude* (or magnitude) of sine wave *1*, shown in Fig. 122*a*, while the 90-degree angle vector *2* makes with vector *1* (arbitrarily drawn horizontally) indicates the 90-degree phase difference between the two sine waves. Fig. 123*d*, finally, illustrates the vector representation of the two 180° out-of-phase sine waves illustrated in Fig. 122*b*. Since vector *1* goes straight up, while vector *2* points down, the opposing directions indicate that the two sine waves tend to cancel each other out.

Rotating Vectors. You must not think of a vector as being necessarily *stationary*. For instance, a point or an object rotating in a circle, such as a stone at the end of a cord, can be represented by a **rotating vector.** The position of the rotating vector *at any particular instant* indicates the distance and direction to the stone (or whatever object) from the center of the circle.

Rotating vectors are very useful to represent the current and voltage variations in an a-c generator. Thus, the rotating radius (in Fig. 120) that generates the a-c sine wave is, of course, a rotating *vec-*

tor. Here we have *frozen* the rotation of the vector at successive instants of time in 30° intervals. In general, all vector diagrams of a-c sine waves assume that the rotating vectors (representing the sine waves) are *frozen at some instant of time.* For example, the vector diagram of the two 90° out-of-phase sine waves in Fig. 122*c* assumes that the phase of the two waves is compared at the particular instant of time when wave *1* is passing through zero, while wave *2* is at its negative maximum or 270°. (See also Fig. 122*a*.) When wave *1* is passing through zero, its vector representation is a *horizontal line to the right,* which is the *conventional* starting point or *reference line* of a counterclockwise rotating a-c vector. (It also corresponds to the horizontal orientation of the rotating radius in Fig. 120 at the start of the cycle, or 0°.) But since wave 2 (in Figs. 122*a* and 123*c*) is at this instant at its negative maximum, having completed 270 degrees of the a-c cycle, it is represented by a vector that has rotated counterclockwise (from the horizontal reference line) through 270° and, hence, points straight down in the page. Fig. 124 should help to clarify the concept of the rotating vector for representing an alternating-current waveform. Here the vector, rotating in a counterclockwise direction has been frozen in four positions, at one-quarter cycle or 90° intervals and, hence, represents the sine wave at right at these instants during the cycle.

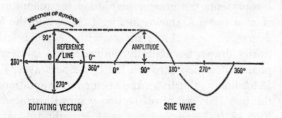

Fig. 124. The Counterclockwise Rotating Vector (at left) Represents the A-C Sine Wave (at Right) at 90-Degree Intervals

Fig. 123*d* is another example that an a-c vector diagram can represent the actual waveforms only at a particular instant of time, with the rotating vectors frozen in position. We could have represented the 180-degree phase difference between the two waves by *any* two lines drawn in opposite directions. The fact that we have drawn vector *1* (representing sine wave 1 in Fig. 122*b*) straight *up* and vector 2 (representing sine wave 2) straight *down* indicates that we are comparing the phase at the instant of time when sine wave *1* reaches its positive maximum or 90° point. As shown in Fig.

122*b*, sine wave 2 is at its negative maximum or 270° point at this instant and, hence, is represented (in Fig. 123*d*) by vector 2, which has rotated through 270° from its horizontal reference line. In general, we are more interested in the phase difference between two a-c quantities than in the instant of time picked to compare their phase. But it helps to keep these facts in mind.

Vector Addition. Vectors, having magnitude and direction, are not as easily dealt with as ordinary numbers. In vector algebra a number of methods are available for calculating with vector quantities; and, particularly, the **algebra of complex numbers** offers a rapid and relatively simple method for handling a-c vectors. These methods are somewhat too advanced for our simplified treatment here and we shall confine ourselves to a graphic explanation of vector addition and subtraction. In the case of *rectangular* vectors (vectors at right angles) a simple numerical treatment is available, which we shall also describe.

Vector "addition" has nothing to do with arithmetic, but refers to the combination of two or more vectors to find their *resultant,* which is also a vector. As an example, assume a man is rowing a boat downstream with a velocity of 4 miles per hour and in a direction parallel to the river banks, as illustrated in Fig. 125*a*. Assume further that a crosswind of 3 miles per hour is pushing the boat horizontally across the river towards the opposite bank.

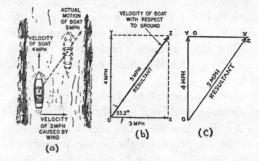

Fig. 125. Vector Addition of Two Velocities; (a), (b), (c)

To find the actual motion of the boat, which is the *resultant* of these two motions, lay off to scale the velocity of the row boat in the *absence* of crosswind as the vector *OY* in Fig. 125*b*. (The velocity of 4 m.p.h. may be represented by 4 inches, 4 cms, or 4 squares of the graph paper, etc.) Then lay off vector *OX* at right angles to *OY*, to represent the velocity of the boat under the influence of the cross wind alone. The two vectors are "added" by completing the parallelogram *OXYZ* and drawing the diagonal, *OZ*, which represents the *resultant or vector sum* of both vectors and shows the actual motion of the boat. If you measure the length of the resultant *OZ* on the same scale as the two component vectors, you will find that it corresponds to a velocity of 5 m.p.h. Measuring the angle between the resultant and the horizontal vector *OX* with a protractor, it will turn out to be 53.2°. The actual motion of the boat, thus, is 5 m.p.h. cross stream at an angle of 53.2° with the horizontal (or 90° − 53.2° = 36.8° with respect to the downstream direction).

Although it may not appear to make much difference, you can obtain the same result without bothering to complete the parallelogram. Simply lay off vector *OX*, without changing its magnitude or direction, next to vector *OY* so that its rear end (*O*) coincides with the arrow point (*Y*) of vector *OY*. (See Fig. 125*c*.) A line drawn from *O* to *Z*, then represents the resultant *OZ*. This toe-to-tip method of vector addition is considerably faster than the parallelogram method when more than two vectors are involved.

As another example consider the vector addition of three a-c voltages that are 45 degrees out of phase with each other. (Fig. 126.) Voltage V_1, which is 200 volts, is represented by horizontal vector OV_1 in Fig. 126*a*. Voltage V_2, 50 volts in magnitude, is represented by vector OV_2, and leads V_1 by a phase angle of 45°. Voltage V_3, 100 volts in magnitude, is represented by vector OV_3, leading V_2 by an angle of 45° or V_1 by an angle of 90°. If these three volt-ages are applied to an a-c circuit, what is the *resultant voltage* acting in the circuit? To answer this question we must find the *vector sum*.

Let us obtain the vector sum of the three voltages first by the parallelogram method, illustrated in Fig. 126*b* and *c*. To do this, you have to proceed one step at a time, finding the resultant of any *two* vectors. In (*b*) of Fig. 126 the parallelogram for vectors $V_1 + V_3$ has been completed and the resultant, V_{13}, is represented by the diagonal, as shown. (It does not matter in which order the vectors are added; V_1 and V_2 could have first been combined equally well.) We do not care to ascertain the magnitude and direction of V_{13}, since it represents only an intermediate result. Next we add vector V_2 to the resultant (V_{13}) of vectors V_1 and V_3, as illustrated in (*c*). Completing the parallelogram of these two vectors and drawing the diagonal, we obtain the resultant, V_{123}, which represents the vector sum of $V_1 + V_2 + V_3$. Measuring the length of this resultant on the scale of the graph paper (i.e., the number of squares), we find the *magnitude* of the resultant a-c voltage V_{123} to be about 272 volts. With a protractor we determine the angle V_{123} forms with the horizontal reference line (or V_1) to be about 30 degrees. The resultant voltage V_{123}, thus, is 272 volts in magnitude and it leads voltage V_1 by a phase angle of 30°, lags voltage V_2 by an angle of 45° − 30° = 15°, and lags V_3 by 90° − 30° = 60°.

Part (*d*) of Fig. 126 shows how this same result

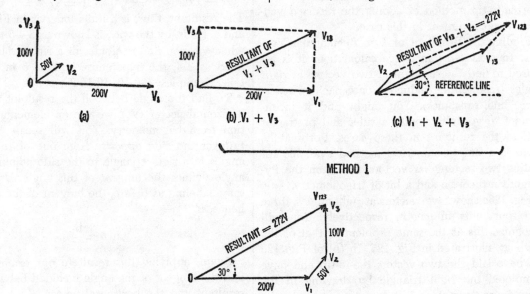

(a)

(b) V₁ + V₃

(c) V₁ + V₂ + V₃

METHOD 1

(d) V₁ + V₂ + V₃ (METHOD 2)

Fig. 126. Vector Addition of Three A-C Voltages

can be obtained much quicker by the toe-to-tip method. Here we have laid off vector V_2 end on to V_1, and vector V_3 at the arrow point of V_2. Connecting the origin (O) to the arrow point of V_3, we get the resultant immediately.

Vector Subtraction. Two vectors may be "subtracted" from each other by *reversing* the vector to be subtracted and then *adding* this reversed vector to the first. For example, in Fig. 127a vector OA is to be subtracted from vector OB. To do this, simply reverse vector OA and add $-OA$ to OB by the parallelogram method. The resultant vector, OC, represents the *vector difference OB minus OA*.

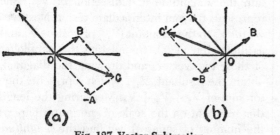

Fig. 127. Vector Subtraction

In contrast, if vector OB is to be subtracted from vector OA (b, Fig. 127), OB is reversed and then $-OB$ is added to OA. The diagonal of the new parallelogram, OC', is the resultant and represents the *vector difference OA minus OB*. If more than two vectors are to be subtracted from each other, you can do this step by step, taking the vector difference of two vectors at a time. You might find the toe-to-tip method of adding the *reversed* vectors more convenient in this case.

Finding the Resultant of Two Rectangular Vectors. In most a-c impedance calculations it is required to find the resultant of two vectors at *right* angles (representing inductive or capacitive *reactances* and resistance). You might find it inconvenient to go around with a ruler and protractor to find the resultant in these cases. Fortunately, there is a simple numerical method available for adding two rectangular vectors, based on the Pythagorean theorem and a bit of trigonometry.

Fig. 128a shows two vectors at right angles, three and four units in length, respectively. You will recognize this as the same problem as that of the rowboat, illustrated in Fig. 125. In (b) of Fig. 128 we have laid the two vectors toe-to-tip and have completed the right triangle by drawing in the hypotenuse (long side). This hypotenuse, obviously, is the *resultant* of the two vectors, corresponding to vector OZ in Fig. 125b and c.

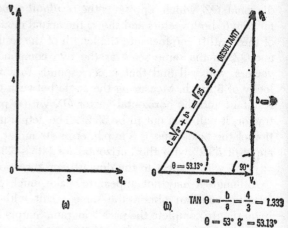

Fig. 128. Finding the Resultant of Two Rectangular Vectors

To find the *length* (magnitude) of the resultant vector, we employ the Pythagorean theorem. According to this theorem, the *square* of the hypotenuse equals the *sum of the squares* of the two sides. If the sides of the triangle are a and b, respectively, and the hypotenuse is labeled c, this may be expressed:

$$c^2 = a^2 + b^2$$

Extracting the square root of both sides of the equation, we obtain for the length of the hypotenuse of a right triangle:

$$c = \sqrt{a^2 + b^2}$$

Applying this result to our example (Fig. 128), we determine the length of the resultant (hypotenuse)

$$c = \sqrt{3^2 + 4^2} = \sqrt{9 + 16} = \sqrt{25} = 5 \text{ units}$$

The resultant, thus, is 5 units long, as in Fig. 125b and c. (This, by the way, is known as 3-4-5 triangle; whenever you find two sides of a right triangle related as 3:4, the hypotenuse will be 5 in proportion. Examples: 6-8-10, 12-16-20, etc.)

To find the *angle* between the resultant and the horizontal vector (V_1), we use an elementary relation from trigonometry. You will recall that the *ratio* of the side *opposite* from one of the acute angles in a right triangle to the side adjoining the angle defines the *tangent* of this angle. Using the same labeling as before, the tangent of the angle θ, hence, is

$$\tan \theta = \frac{b}{a}$$

Again, applying this result to our example, we find the tangent of the angle included between the resultant and the horizontal vector

$$\tan \theta = \frac{4}{3} = 1.333$$

You can find the angle corresponding to this tangent in any book of tables of the trigonometric functions. The angle θ turns out to be 53.13° (53°8′), or approximately the same as we found graphically in Fig. 125b.

Resolving the Resultant into Two Rectangular Component Vectors. Let us now look for a moment at the reverse problem: given the *resultant* or vector sum of two vectors, can we find the *component vectors* that add up to the given resultant? This problem might arise, for instance, when the motion of an object is known to be caused by two forces and it is desired to find the magnitude and direction of these *component* forces.

Fig. 129 shows the vector representation of this problem. All we have is the magnitude and direction of the resultant vector, as illustrated in (a). Since the resultant represents the vector sum of *two* component vectors, we know immediately that it must be the hypotenuse (long side) of a triangle, whose short sides are made up by the two component vectors. This does not solve the problem, however, since we can draw any number of triangles with the same hypotenuse (resultant), as illustrated in Fig. 129b. To *resolve* the resultant into its component vectors, we must know the *angle* between the component vectors, so that their relative directions are fixed. Given the resultant and the angle at which the components are acting, only *one* vector triangle may be drawn with two definite component vectors. The dotted triangles of Fig. 129b all have the same resultant and represent various graphical solutions of the problem depending on the angle between the components.

Fortunately, most a-c problems require that a given resultant (such as **impedance**) be resolved into two component vectors at *right* angles (such as **reactance and resistance**). For two *rectangular* component vectors, the problem is not only solved quickly by graphical means, as shown in Fig. 129c, but also lends itself to a simple numerical method. In Fig. 129a the resultant (side c of a vector triangle) has a length of 13 units and it is inclined at an angle of 22.6° (22°35′) with respect to the horizontal of the page. You will recall from elementary trigonometry that the *ratio* of the side *adjacent* (a) to this angle to the hypotenuse (c) defines the *cosine* of the included angle (θ) and the ratio of the opposite side (b) to the hypotenuse defines the *sine* of the included angle θ. Hence, we can write

$$\frac{a}{c} = \cos\theta, \text{ or } a = c\cos\theta; \text{ also } \frac{b}{c} = \sin\theta,$$

or $b = c\sin\theta$

In our example, the resultant $c = 13$ units and $\theta = 22.6°$. Thus, the horizontal vector V_1 (side a) = 13 cos 22.6° = 13 × 0.9233 = 12 units and the vertical vector V_2 (side b) = 13 sin 22.6° = 13 × 0.384 = 5 units. (The values of sine and cosine are from tables.)

Practice Exercise No. 11

1. Define the following terms: *alternation, cycle, frequency period, amplitude, peak voltage or current, angular velocity.*

2. (a) How many radians are there in 360 "electrical" degrees? (b) the following readings were taken from the time scales (X-axis) of various sine waves: 180°, $\pi/4$ radians, 360°, 4π radians, 30°, $\pi/3$ rad., 720°, $2\pi/3$ rad., 270°, $3\pi/2$ radians. Translate these values into cycles or fractions of a cycle.

3. A 2-pole generator rotates at 3000 rpm and produces a peak sine voltage of 115 volts a.c. What is the frequency? Also, write the expression for the instantaneous voltage at any time.

4. The peak value of a 25-cycle alternating current is 30 amps. (a) What is the effective value of this current; (b) what is its instantaneous value 0.002 sec. after passing through zero (in a positive direction)?

5. Distinguish between *peak* (maximum), *effective* (rms) and the *average* value of an alternating current and give the mathematical relationships.

6. An a-c ammeter reads 22 amps (rms) current through a load and a voltmeter reads 385 volts (rms) drop across the load. What are (a) the peak values and (b) the average values of the alternating current and voltage?

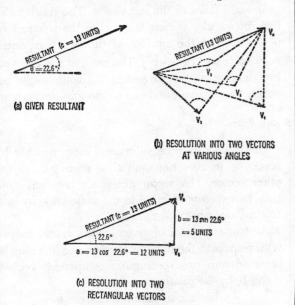

(a) GIVEN RESULTANT

(b) RESOLUTION INTO TWO VECTORS AT VARIOUS ANGLES

(c) RESOLUTION INTO TWO RECTANGULAR VECTORS

Fig. 129. Resolving a Resultant into Component Vectors

7. Sine wave *1* leads sine wave *2* by a phase angle of 60° and lags sine wave 3 by 130°; what is the angle between wave *2* and wave 3? Which is leading?

8. Distinguish between *scalar* and *vector* quantities. Can you add and subtract vectors like scalar quantities?

9. An airplane flying due north at 360 mph is subjected to a cross wind due west of 150 miles per hour. Draw a vector diagram to represent the situation and determine the *resultant* velocity of the airplane relative to the ground.

10. Vector *A* is 200 units long and forms a right angle with vector *B*, which is 150 units long. (a) What is the length of the resultant of the vector sum *A* + *B* and what angle does it make with vector *B*? (b) Draw diagrams of the vector *differences A − B* and *B − A*. Is the *magnitude* of the resultant the same for vector subtraction as for vector addition in this case?

11. The total *impedance* of an a-c circuit is 500 ohms and has a phase angle of 30 degrees with respect to the horizontal (representing the applied voltage). If the impedance represents the vector *sum* of a vertical *reactance* vector and a horizontal *resistance* vector, find the values of the reactance and the resistance in the circuit.

SUMMARY

One a-c sine-wave cycle, consisting of a positive and a negative **alternation**, contains 360 electrical degrees or 2π radians.

The maximum value or crest of each sine-wave alternation, called its **amplitude**, determines the **peak** value of voltage or current.

The number of cycles completed per second (cps) is called the frequency of the sine wave, and the time required to complete one full cycle is the period of the wave. ($T = 1/f$; $f = 1/T$.)

Angular velocity (in radians) is 2π times frequency ($w = 2\pi f$). The product of angular velocity and time (wt) equals the **angular displacement**, angle of rotation, or **phase angle** (θ) of the sine wave.

The instantaneous values of a sinusoidal alternating current (i) or voltage (e) may be represented by the relations

$$i = I_m \sin \theta = I_m \sin wt$$
and $$e = E_m \sin \theta = E_m \sin wt$$

where I_m and E_m are the *peak* values of the current and voltage, $\theta = wt$ is the angular displacement (phase angle), $w = 2\pi f$ is the angular velocity, t = time, and f = frequency.

The relations between the effective (root-mean-square), peak and average values of an alternating current and voltage are:

$$I_{rms} = 0.707 \, I_m; \, I_{av} = 0.636 \, I_m = 0.9 \, I_{rms}$$
$$E_{rms} = 0.707 \, E_m; \, E_{av} = 0.636 \, E_m = 0.9 \, E_{rms}$$
and $$I_m = 1.414 \, I_{rms} = 1.57 \, I_{av}$$
$$E_m = 1.414 \, E_{rms} = 1.57 \, E_{av}$$

The **phase angle** of an a-c voltage or current is the **fraction** of the cycle that has elapsed from the (zero) starting point, measured in electrical degrees or radians. **Relative phase, phase angle** or **phase difference** (in fractions of a cycle or electrical degrees) are also used to compare two a-c voltages, currents, or a voltage and a current of the *same* frequency that pass through their respective zero values at different instants of time.

A **vector quantity**, having *both* magnitude and direction, may be represented by a straight-line segment, whose **length** denotes the **magnitude** of the vector and whose orientation with respect to a reference line denotes the **direction** of the vector quantity.

The current and voltage variations in an a-c generator may be represented by **rotating vectors**, whose instantaneous positions represent the magnitude and phase of the current or voltage at that moment. A-C vector diagrams are **rotating vectors frozen in time**.

Two or more vectors may be **added** successively by completing the parallelogram of any two vectors and drawing in the diagonal, representing the resultant or **vector sum**; they may also be added by placing the vectors toe to tip and drawing the resultant from the starting point of the first vector to the end point of the last vector. The resultant of two **rectangular** vectors is equal to the **square root of the sum of the squares of the two component vectors**: ($c = \sqrt{a^2 + b^2}$). The **ratio** of the two vectors determines the **tangent** of the acute angle between the resultant and one of the adjacent vectors

$$(\tan \theta = \frac{a}{b}).$$

One vector may be **subtracted** from another by reversing its direction and then adding it to the other vector. The **vector difference** between more than two vectors is found in the same way, by subtracting one vector at the time.

A **resultant vector** of given length, *c*, and angle θ with respect to the horizontal reference line may be resolved into two rectangular component vectors, *a* and *b*, by the relations:

horizontal vector, $a = c \cos \theta$
vertical vector, $b = c \sin \theta$

CHAPTER TWELVE

ALTERNATING-CURRENT CIRCUIT COMPONENTS

Did you ever confront an unfamiliar power outlet and wonder whether it supplied a.c. or d.c.? There is nothing in the nature of the current flowing through the light bulbs in the room that would tell whether they were powered by a.c. or d.c. If you plugged in your electric shaver, it would certainly work, since it is designed to operate on either a.c. or d.c. If you should happen to touch both exposed poles of the power outlet, you would get a very unpleasant shock, whether it was a.c. or d.c., and we strongly advise against this useless, dangerous, and possibly lethal test. Just what can you do to solve this riddle?

There are several tests you can perform to check the type of current supply, provided you did some shopping first. Probably the simplest is a chemical test with litmus paper. If the positive pole of a d-c supply is touched to *wet litmus paper*, the paper will turn red. Thus, if you get a red reaction from one of the poles, you have d.c.; if not, the supply is a.c. Another relatively simple test would be with a *neon glow bulb* (which is available everywhere in the form of translucent plastic plugs, and which serves as a night light). When plugged into a d-c supply only *one* of the poles (the *negative* electrode) will glow with a characteristic orange light; when plugged into a.c., in contrast, *both* electrodes will glow. If you run into this problem often, it may be handy to have a neon test light with you at all times.

If you are scientifically-minded, the simple tests with litmus paper or a neon bulb will probably not satisfy you and you will want to investigate the behavior of the current supply in an actual *circuit*. To do this you should get a hold of an ac-dc volt-ammeter (or separate meters), several resistors of various values, a few capacitors and an inductor in the form of a choke coil or transformer. Now you're ready to experiment (See Fig. 130.)

You can start by hooking several resistors together or selecting one sufficiently large and with sufficient power rating to be safely connected across a 100-200 volt power line. With the voltmeter, you first measure the applied voltage by connecting the meter directly across the line. Then you measure the current through the simple resistive circuit, by connecting an ammeter *in series* with the resistors, as shown in Fig. 130a. If now, in accordance with Ohm's Law, you divide the voltage reading by the current reading, you will obtain the value of the resistance or resistors you have selected ($R = E/I$). This is true regardless of whether the supply is a.c. or d.c. Hence, a resistance *alone* cannot tell you whether you have alternating or direct current.

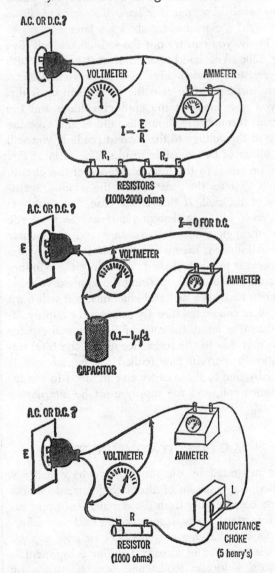

Fig. 130. A-C and D-C Behavior of Resistance, Capacitance and Inductance. (a), (b), and (c)

129

Next, you're ready to try a capacitor. You can connect it in series with a resistor to limit the current or directly across the line, if its voltage rating exceeds that of the power source (See Fig. 130*b*.) The voltmeter will read exactly the same applied voltage as before. The ammeter, connected in series with the capacitor, may indicate a certain current reading or it may read *nothing at all*. If it reads nothing at all, the capacitor is connected to a d-c source; if a current is indicated, the capacitor is connected to an a-c source. A capacitor thus presents an *open circuit* to direct current, as we should have expected. Since the plates of a capacitor are not connected to each other, no *direct current* should flow after the initial brief charging pulse. How and how much *alternating current* flows "through" a capacitor we shall see later.

Finally, you can try out the inductance coil (See Fig. 130*c*.) You had better connect it in series with the resistors used in Fig. 130*a* to prevent the coil from burning out in case the power supply is d.c. If you now measure the applied voltage and the current through the circuit, as before, and take the *ratio* of the voltage to the current reading, you will get either of two possible results. The ratio of E/I may be equal to the total resistance of the circuit, which includes the resistors and the *winding resistance* of the coil. If that is the case, the supply is d.c., since for an unchanging current, the magnetic field through the coil is constant and no counter emf is induced, except for the brief instant when the field is first established. With only the winding resistance of the coil opposing the applied voltage, you can easily see that an inductance coil will burn out when connected directly across a d-c supply. If, on the other hand, the ratio of E/I is much greater than that due to the resistance alone, the total opposition to current flow (called the *impedance*) is partially due to the counter emf induced in the inductance coil, and the supply must be *alternating current*.

A-C CIRCUIT COMPONENTS

As suggested by the problem we have just described, the behavior of alternating-current circuits differs considerably from d-c circuits—coils and capacitors, particularly, responding in radically different ways to varying currents. Let us now explore the a-c behavior of already familiar components—resistors, inductors (coils) and capacitors—one by one, before going into their more complicated combined behavior in a-c circuits. To do this we have to make use of an abstraction—so-called "pure" components. By assuming "pure" resistors, inductors, and capacitors we shall be able to analyze their a-c behavior separately. Let us say at the outset that this procedure is at best an approximation. There are no pure components. A wire-wound resistor, for example, not only has a certain resistance, but also a certain inductance, which depends on the length of the wire, the number of turns, etc.; and in addition a certain capacitance, which depends on the spacing between turns, the number of turns, the "dielectric" (insulating material) and other factors. Similarly, any inductance coil wound of wire not only has inductance, but also the total resistance of its winding and a certain capacitance between individual turns. Even a capacitor, which almost attains the ideal of "purity" has the resistance and inductance of its terminal leads and, moreover, (because of its imperfect dielectric) behaves as if a high resistance were shunted across it through which the charge on its plates slowly "leaks off."

At high a-c frequencies, particularly, many resistors begin to "look like" and behave like inductance coils, coils start to behave like capacitors, and capacitors like coils. By special tricks, such as winding a resistor "non-inductively" (by doubling up the winding so that the current flows in opposite directions through adjacent turns), the components may be made fairly pure even at high frequencies. At the low a-c power frequencies we shall be concerned with (50 or 60 cycles in most countries), we may neglect these "impurities" with very little error.

RESISTANCE IN A-C CIRCUITS

Fortunately, the average resistor cares little whether direct or alternating current circulates through it. Except at high frequencies (above many thousands of cycles), the **a-c resistance of a given conductor or resistor is the same as its d-c resistance.** Thus, if a resistor of a given value R is connected either to a d-c or a-c voltage E, the current immediately assumes the value E/R, predicted by Ohm's Law. For a.c. usually the *effective* (rms) value of the voltage (E) is stated, in which case the application of Ohm's Law will give the effective (rms) value of the current (I = E/R). If we plot the relationship between a sine-wave a-c voltage (E) applied to a resistor and the resulting current (I) through it, we obtain the waveform illustrated in Fig. 131. The voltage and current are *in phase*, going through their respective maxima and minima together, and the *ratio* of their magnitudes (E/I) at any point is equal to the resistance.

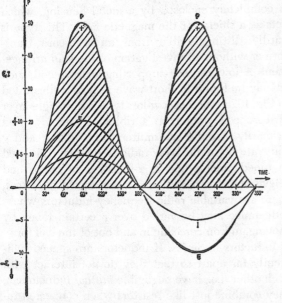

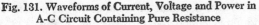

Fig. 131. Waveforms of Current, Voltage and Power in A-C Circuit Containing Pure Resistance

The power consumed in a resistive a-c circuit is simply the product of the effective values of voltage and current (P = E × I), or the *square of the current times the resistance* (P = I² R), or the *square of the voltage divided by the resistance* (P = $\frac{E^2}{R}$), just as for d.c. Fig. 131 also illustrates the power graph of a resistive a-c circuit, obtained by multiplying together the instantaneous voltage (*E*) and current (*I*) (ordinates) at points along the time axis. The shape of this curve is a sine-square (sin²) wave, with *both half-cycles positive*, as we have already seen (See Fig. 121.) The positive half-cycles of the power graph indicate that power is *consumed* at all times in a resistive a-c circuit.

EXAMPLE: An ammeter indicates a current of 2 amperes through a resistor that is connected to 120 volts *d.c.* The resistor is then connected to an *a-c* voltage with a *peak* value of 170 volts. What is the a-c current through the resistor and the power consumed in the resistance?

Solution: The d-c or a-c resistance R = $\frac{E}{I} = \frac{120}{2}$ = 60 ohms.

The rms (effective) value of the a-c voltage =

0.707 × 170 = 120 volts, or the same as the d-c voltage. Hence the (rms) current is again I = $\frac{E}{R}$.

= $\frac{120}{60}$ = 2 amperes. The power consumed

P = E × I = 120 × 2 = 240 watts;
or P = I² R = (2)² × 60 = 4 × 60 = 240 watts;
or P = E²/R = (120)²/60 = 14400/60 = 240 watts.

Skin Effect. Although the d-c and a-c resistance of conductors is the same at low (power) frequencies, this is no longer true at frequencies in the order of thousands of millions of cycles. At these frequencies the current has a tendency to flow near the *surface* of a conductor, a phenomenon known as the **skin effect**. Because of the skin effect, a high-frequency current does not make use of the complete cross section of a conductor, but only of a small portion of it (the "skin"), and the a-c resistance goes up proportionately. The greater the diameter of a conductor, the more pronounced is the skin effect; that is, the less of the conductor's cross section is utilized. The skin effect and, hence, the a-c resistance, also goes up as the *square root of the frequency*; as a consequence, very little of the cross section of a conductor is utilized at extremely high frequencies. This has led to the use of **tubular conductors** at very high frequencies to eliminate the unnecessary central portion of the conductor. In addition, high-frequency conductors for important applications (radar, for example) are frequently silver-plated on the outside to reduce the resistance of the surface layer.

As an example, let us see what happens to the resistance of 1000 feet of No. 10 copper wire, as the frequency of an alternating current goes up. You will recall that No. 10 copper wire has a diameter of about 0.1 in. (0.102 in. exactly) and 1000 feet of it have a *d-c* resistance of about 1 ohm. If an alternating current with a frequency of 10 kilocycles (10,000 cps) is passed through the wire, it will penetrate the conductor only to a depth of about 0.03 in. (30 mils) and the resistance of 1000 feet will increase to about 1.25 ohms. If the frequency of the current is increased to 1 megacycle (1,000,000 cps), the current will penetrate the wire to a depth of only about 0.003 in. (3 mils) and the resistance goes up to about *10 ohms* per 1000 feet. Finally, increasing the frequency to a 100 mc (10⁸ cps) will decrease the "skin depth" to about 0.0003 in. (0.3 mil) and increase the resistance of 1000 feet of wire to about 98 ohms.

INDUCTANCE IN A-C CIRCUITS

We have already discussed the property of inductance in conductors and coils which by producing a counter emf opposes any *change* in the current flowing through the inductor. We have also seen that the magnitude of this counter emf is equal to the product of the inductance and the *rate of change* of the current through the coil ($E = -L\frac{\Delta i}{\Delta t}$). Up to now we have considered inductance primarily in connection with direct or slowly varying currents. The real importance of inductors, however, lies in their application to a-c circuits.

A typical assortment of inductors for a-c circuits is illustrated in Fig. 132. The construction and design of an inductor depends, of course, on the application. We distinguish primarily between **iron-core inductors** (called **chokes**), used for power and

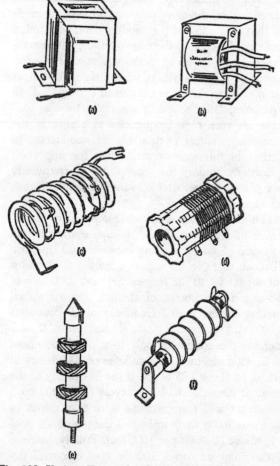

Fig. 132. Various Types of Inductors: (a) and (b) iron-core chokes; (c) low-loss air-core, self-supporting coil; (d) single-layer air-core coil on ribbed frame; (e), (f) and (g) multi-layer radio-frequency coils and chokes

filtering applications at low frequencies, and a **variety of air-core inductors**, used in electronic applications at high (radio) frequencies. Fig. 132*a* and *b* shows two typical iron-core chokes; the choke in (*a*) is completely enclosed by a metal housing, which acts as a **shield** for the magnetic field. This type is hardly distinguishable from an iron-core transformer without an investigation of the lead arrangement. A low-loss, self-supporting air-core coil, such as may be found in short-wave radios, is illustrated in Fig. 132*c*. Another low-loss type is the single-layer air-core coil wound on a ribbed frame (*d*), used primarily in radio transmitters. Fig. 132*e*, *f* and *g* illustrate the multi-layer radio-frequency coils used in tuned circuits and as r-f chokes in radios and television sets. In addition to these **fixed inductors**, there are **variable radio-frequency inductors** whose inductance may be varied over a certain range by moving an iron-core slug in and out of the coil form.

Inductors in Series. If inductors are spaced sufficiently far apart so that they do not interact with each other (i.e., have negligible *mutual* inductance), they combine just like resistors when connected together. Thus, the total inductance (*L*) of a number of inductors connected in series, as in Fig. 133, is simply the *sum* of the individual inductances, or expressed mathematically

$$L = L_1 + L_2 + L_3 + L_4 + \ldots$$

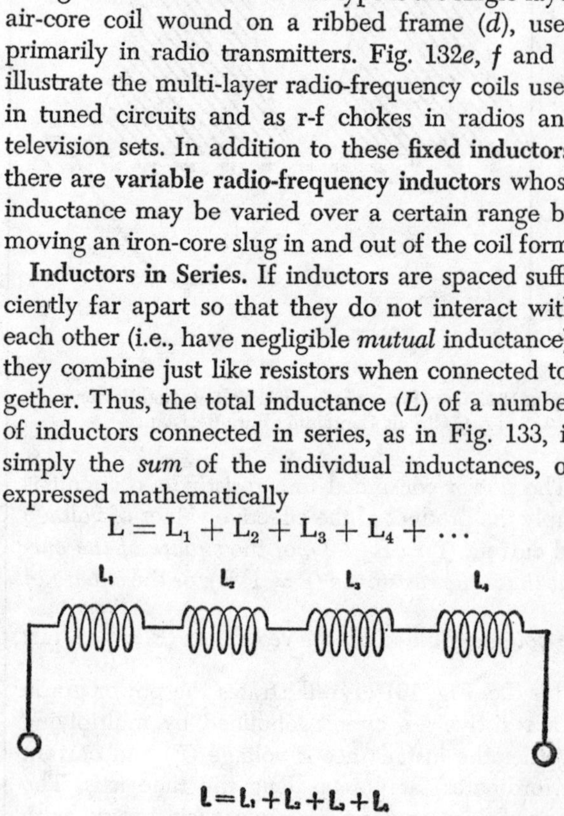

Fig. 133. Inductors in Series

But if two series-connected coils are spaced close together so that their magnetic field lines interlink, their total inductance is

$$L = L_1 + L_2 \pm 2M$$

where *M* represents the *mutual* inductance between the coils (in henrys). The plus sign in the expression above is used, if the coils are arranged in series-aiding (Fig. 134*a*); that is, in such a manner that the current flows through the turns of both coils in the same direction and the magnetic fields assist each other. The minus sign is used if the coils are connected in series-opposing; that is, in such a manner that their magnetic fields oppose each other (Fig. 134*b*).

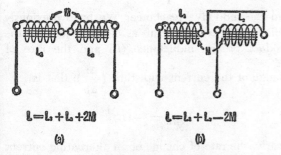

$$L = L_1 + L_2 + 2M$$

(a)

$$L = L_1 + L_2 - 2M$$

(b)

Fig. 134. Mutually Coupled Inductors Connected in Series-Aiding (a) and in Series-Opposing (b)

EXAMPLE: A 5-henry and a 12-henry choke are connected in series and, initially, spaced far apart. What is their total inductance? The coils are then moved close together so that they are coupled by a mutual inductance of 7 henrys. What is their total inductance, if (a) the current flows through the coils in the same direction and (b) the current flows through the coils in opposing directions?

Solution: The inductance of the two *uncoupled* coils in series is

$$L = L_1 + L_2 = 5 + 12 = 17 \text{ henrys}$$

When the mutual inductance is 7 henrys and the coils are connected in series-aiding, $L = L_1 + L_2 + 2M = 5 + 12 + 14 = 31$ henrys; when the coils are connected in series-opposing, their total inductance $L = L_1 + L_2 - 2M = 5 + 12 - 14 = 3$ henrys.

Inductors in Parallel. When inductance coils are spaced sufficiently far apart so that their mutual inductance can be neglected, the rules for combining inductors in parallel are the same as for resistors. Thus, as for two resistors in parallel, the total inductance of *two* coils in parallel (Fig. 135a) is

$$L = \frac{L_1 \times L_2}{L_1 + L_2}$$

Again as for resistors in parallel, the total inductance of a number of coils in parallel is equal to the reciprocal of the sum of the reciprocals of the individual inductances (Fig. 135b). Expressed as a formula,

$$L = \frac{1}{\dfrac{1}{L_1} + \dfrac{1}{L_2} + \dfrac{1}{L_3} + \dfrac{1}{L_4} + \cdots}$$

EXAMPLE: A number of inductors are connected together into an inductance network. Group A is made up of four 12-henry chokes in parallel; group B of a 3-henry and a 5-henry choke in parallel; group C of a 4-henry and a 6-henry choke in paral-

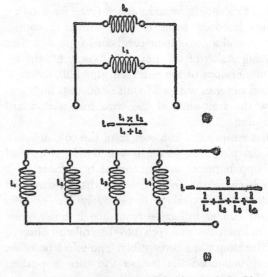

$$L = \frac{L_1 \times L_2}{L_1 + L_2}$$

$$L = \frac{1}{\dfrac{1}{L_1} + \dfrac{1}{L_2} + \dfrac{1}{L_3} + \dfrac{1}{L_4}}$$

(b)

Fig. 135. Inductors in Parallel: (a) Two Coils; (b) Several Coils

lel; and group D of a 7-henry and a 9-henry choke in parallel. Groups A, B, C, and D are then connected in series. What is the total inductance?

Solution: The inductance of group A,

$$L_A = \frac{1}{4/12} = \frac{12}{4} = 3 \text{ henrys}$$

The inductance of group B,

$$L_B = \frac{3 \times 5}{3 + 5} = \frac{15}{8} = 1.875 \text{ henrys}$$

The inductance of group C,

$$L_C = \frac{4 \times 6}{4 + 6} = \frac{24}{10} = 2.4 \text{ henrys}$$

The inductance of group D,

$$L_D = \frac{7 \times 9}{7 + 9} = \frac{63}{16} = 3.94 \text{ henrys}$$

Hence, the total inductance of all groups connected in *series,*

$$L = L_A + L_B + L_C + L_D = 3 + 1.875 + 2.4 + 3.94 = 11.22 \text{ henrys.}$$

Inductive Reactance. We have noted that an inductance opposes any change in the magnitude of a current flowing through it. Since an alternating current, by definition, is changing continuously, an inductance has a *constant* opposition to it, termed **inductive reactance.** You can easily convince yourself of this reactance by making the following simple experiment.

EXPERIMENT 23: Obtain a choke coil of several henrys inductance with a removable iron core (called a **solenoid**), or make one yourself by winding several hundred turns of fine, insulated copper wire around a bakelite or cardboard form and then inserting a slug of soft-iron that loosely fits the inside dimensions of the coil (see Fig. 136). Connect the coil in series with a 25-watt or 40-watt bulb and screw the free ends of the wire into a standard male plug.

First remove the iron core from the coil and connect the plug into an *a-c* power outlet. Note that the lamp burns at almost normal brightness. This is so because the inductance of the air-core coil (without the iron slug) is relatively low and its counter emf or inductive reactance is small. Now start to insert the iron core into the coil and observe that the lamp dims perceptibly. The effect becomes more pronounced the farther the core is pushed into the coil, and the lamp may be almost extinguished when the slug is completely inserted. What is happening is that the magnetic field inside the coil becomes greatly strengthened by the insertion of the high-permeability iron core and, hence, the inductance of the choke coil is tremendously increased. This, in turn, multiplies the counter emf or inductive reactance to the current by the same factor and results in weakening the current through the bulb. As long as the core is inserted, there is a *continuous choking effect on the lamp*. If you repeat the same experiment with *d.c.*, you will observe only a momentary dimming of the lamp when the iron slug is first inserted.

Magnitude of Inductive Reactance. What is the relation between the magnitude of the counter emf and the inductive reactance? We have previously defined the counter emf as being equal to the *product* of the inductance (L) and the rate of change of the current with time ($\frac{\Delta i}{\Delta t}$); that is

$$E_{cemf} = -L\frac{\Delta i}{\Delta t}$$

clearly, the rate of change of an alternating current must be *proportional to its frequency*, since more a-c cycles are completed in a given time as the frequency increases. In fact, it is easily shown in elementary calculus that for an alternating current of instantaneous value $i = I_m \sin wt$, the counter emf

$$E = -L\frac{di}{dt} = -wL\,I_m \cos wt$$

(where di/dt is the calculus notation for the rate of change of current with time, or the *derivative*.) The first part of the right-hand expression above, wL (disregarding the sign), refers to the **inductive reactance**, which has the symbol X_L. (Reactance, in general is denoted by the symbol X and is measured in ohms.) Here w stands, of course, for the angular velocity and is equal to $2\pi \times$ frequency. Thus, we have the simple formula for inductive reactance

$$X_L = wL = 2\pi fL = 6.283\,f\,L \text{ (ohms)}$$

where f is the frequency and L is the inductance.

The second part of the right-hand expression above, $I_m \cos wt$, is the same in *magnitude* as the original current i ($= I_m \sin wt$), since a cosine wave is identical with a sine wave, except that it is *displaced in phase by 90 degrees* (¼ cycle). Thus, we

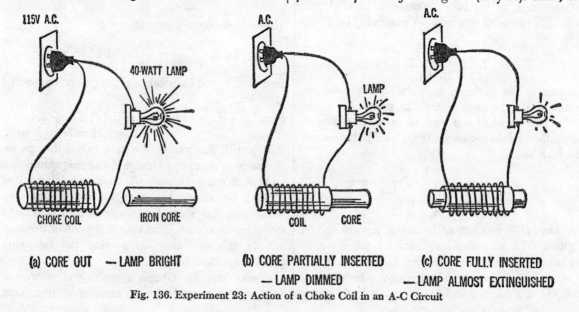

(a) CORE OUT — LAMP BRIGHT   **(b) CORE PARTIALLY INSERTED**   **(c) CORE FULLY INSERTED**
                                                  — LAMP DIMMED   **— LAMP ALMOST EXTINGUISHED**

Fig. 136. Experiment 23: Action of a Choke Coil in an A-C Circuit

see that the **magnitude of the counter emf is simply** the product of the inductive reactance and the original current. (We shall soon consider the phase relations between current and voltage.)

EXAMPLE: A 10-ampere (peak value), 60-cycle, a-c current flows through a 20-henry choke. Write the expression for the instantaneous current and the counter emf, and compute the values of the inductive reactance and the magnitude of the counter emf.

Solution: the angular velocity $w = 2\pi f = 6.283 \times 60 = 377$. Hence, the instantaneous current $i = I_m \sin wt = 10 \sin 377 t$. The inductive reactance $X_L = wL = 377 \times 20 = 7540$ ohms. The instantaneous value of the counter emf

$E = -wL\, I_m \cos wt = -7540 \times 10 \cos 377 t = -75,400 \cos 377 t$. The *peak* value of the counter emf, thus, is 75,400 volts. The effective (rms) value of the counter emf is, therefore,

$$0.707 \times 75,400 = 53,400 \text{ volts.}$$

Magnitude of Current in Inductive Circuit. We have to modify Ohm's Law only slightly to compute the value of the current in an inductive a-c circuit. Since, as we have just seen, the total opposition to an alternating current in a pure inductance is given by the inductive reactance, X_L, the **current is simply the impressed voltage (V) divided by the inductive reactance** (X_L). Thus, we have the simple relation for the magnitude of the current in an inductive circuit:

$$I = \frac{V}{X_L} = \frac{V}{2\pi fL} = \frac{V}{6.283\, fL}$$

EXAMPLE: What is the magnitude of the (rms) current flowing through a 5-henry choke connected across the 115-volt, 60-cycle, a-c line?

Solution: $I = \dfrac{V}{2\pi fL}$

$$= \frac{115}{6.283 \times 60 \times 5} = \frac{115}{1885} = 0.061 \text{ ampere.}$$

Phase Relations. Fig. 137a illustrates the circuit of a pure inductance L, connected across an applied a-c voltage, V. The inductive reactance of the coil permits just enough current to flow to induce a counter emf, E, that *equals and opposes the applied emf.* Thus,

$$V = I X_L = -E$$

(since $E = -wL\, I_m \cos wt = -X_L\, I$). Moreover, we have seen that there is a 90-degree phase difference between the current and the counter emf. The resulting phase relations between the current

(I), the applied voltage (V) and the counter emf (E) are pictured by the instantaneous waveforms of Fig. 137b and the vector diagram (c).

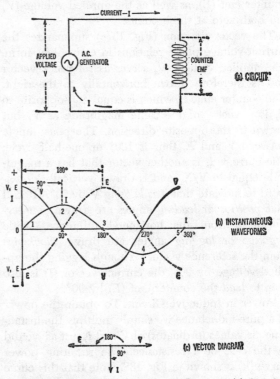

Fig. 137. Phase Relations in a Pure Inductance—(a) Circuit; (b) Instantaneous Waveforms; (c) Vector Diagram

Since the counter emf, E, is everywhere equal and opposite to the applied voltage, V, the two waveforms are 180 degrees out of phase, as is evident in Fig. 137b. Moreover, the current (I) is seen to *lag behind* the impressed voltage (V) by a *phase angle of 90° or one-quarter cycle.* Since V and E are 180° out of phase, the current also *leads* the counter emf in phase by 90° or one-quarter cycle. It is easily seen from the waveforms that this must be so. The magnitude of the counter emf is proportional to the *rate of change* of the current through the coil. It is evident that the rate of change is given by the *slope* of the tangent at any particular point of the instantaneous current waveform. At points 1, 3, and 5, where the current wave crosses the zero line (X-axis), the *rate of change* of the current (slope) is a *maximum* and, hence, the magnitude of the induced counter emf (E) is also a *maximum.* (Equivalently, since V is 180° out of phase with E, the applied voltage is also at a maximum at the zero current points, but opposite in polarity from the counter emf.) In contrast, at points 2 and 4, where the current wave goes through its maximum points, the rate of change or *slope* of the waveform is instantaneously *zero* and, hence, the induced counter emf (*E*), as well as the applied voltage (*V*) are both zero at these points.

The vector diagram (Fig. 137*c*) summarizes the current-voltage phase relations in convenient form. The applied voltage, *V*, serves as reference vector and is therefore drawn horizontally to the right. The counter emf, *E*, which is equal and opposite to *V*, is a vector of the same magnitude as *V*, but drawn in the opposite direction. The phase angle between *V* and *E*, thus is 180° or one-half cycle. The current, *I*, is another vector that has a magnitude equal to V/X_L and is drawn vertically *downward* to indicate that it is *lagging V* by 90°. (Recall that *positive* or *leading* angles are drawn *counterclockwise* from the horizontal reference, while *negative* or *lagging* angles are drawn *clockwise* from the reference vector.) Though lagging the applied voltage by 90°, the current vector (*I*) is also seen to *lead* the counter emf (*E*) by 90°.

Power in Inductive Circuit. To obtain the power in a pure inductance we simply multiply the instantaneous values (ordinates) of *V* and *I*, just as we did in the case of a resistance. The resulting power curve (*P*) is shown in Fig. 138. Note that this curve has *two positive and two negative loops during one*

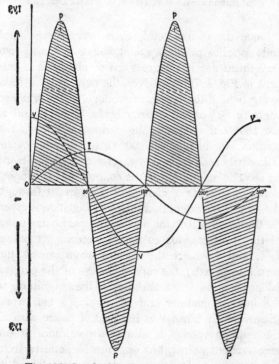

Fig. 138. Graph of Power in a Pure Inductance

complete cycle of the impressed voltage (*V*). Since *positive* power means power *consumed* or absorbed *by the circuit*, while *negative* power indicates power *returned from the circuit*, it is evident that **no net or real power is consumed by a pure inductance**. During one-quarter cycle energy is *supplied to* the inductor and in the next quarter-cycle the same amount of energy is *returned by* the inductor. Physically this simply means that the energy contained in the magnetic field, when it is built up by current flow in one direction, is returned by the collapsing magnetic field when the direction of current flow reverses. Although a pure inductance does not absorb any power, you must remember that all actual (physically realizable) inductors always have some resistance and hence always *absorb some amount of power*. Just how much, we shall see presently.

CAPACITANCE IN A-C CIRCUITS

We have already considered the action of two metal plates separated by a dielectric—a capacitor. We saw then that the plates of a capacitor rapidly charge up to the voltage of the source when it is connected to a d-c supply. The energy taken from the source is *stored in the electrostatic field* between the plates. Nothing further happens after the brief charging pulse, and the current ceases (because of the electrostatic repulsion between the charges) when the potential between the plates attains the same value as that of the source. A capacitor, therefore, is an effective **barrier to direct current**.

When a capacitor is connected to a source of *alternating current*, the plates become charged, then discharged, and charged again in the opposite direction, in rapid sequence and in accordance with the alternating polarity of the applied voltage. As a consequence, the electrons surge to and fro in the connecting wires of the capacitor and give rise to an alternating current that is said to flow "through" the capacitor. Because of the insulating dielectric no current can actually flow "through" the capacitor, but since a current surges back and forth in the connecting wires, the effect is the same as if an alternating current would flow through it. A capacitor, therefore, does *not* bar the flow of alternating current, though it materially *reduces* it. You can convince yourself of this basic fact by the following experiment.

EXPERIMENT 24: Procure a 2-5 microfarad, 150-

volt *paper* capacitor. (Do not buy the *electrolytic* type, since it is useless on a pure a-c source.) Connect the capacitor *in series* with a 25-40 watt bulb and plug the free ends of the wire into the 115-volt a-c line (See Fig. 139). Note that the lamp lights, though much more dimly than when it is connected directly across the a-c line. If you have a source of 115 volts d.c. available, you can repeat the experiment and convince yourself that the lamp will not burn at all, since the capacitor blocks d.c.

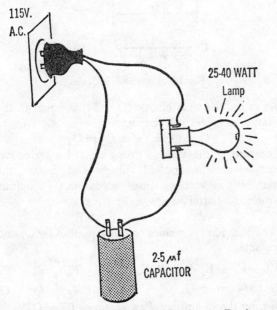

Fig. 139. Experiment 24: A Capacitor is not a Barrier to Alternating Current

Types of Capacitors. Capacitors come in various types, shapes and sizes depending on use and the requirements of the particular circuit. An assortment of some typical capacitors, used primarily in electronic applications, is illustrated in Fig. 140. Note the variety of shapes.

Capacitors are frequently used in radio and other electronic circuits to by-pass (shunt) alternating currents of relatively high frequency when direct current is also present. When the frequency is high and losses are to be kept low, fixed **mica capacitors** are used, which use flat sheets of relatively expensive mica as dielectric between the plates. (Fig. 140a and *b*.) At low frequencies and where losses are not critical, **paper by-pass capacitors** are used (*c* and *d*). These are made of alternate sheets of wax-impregnated paper and metal foil, which are rolled or folded for compactness. For filtering residual a.c. in a d-c power supply, large **electrolytic filter capacitors** are used (*e*). The latter are made

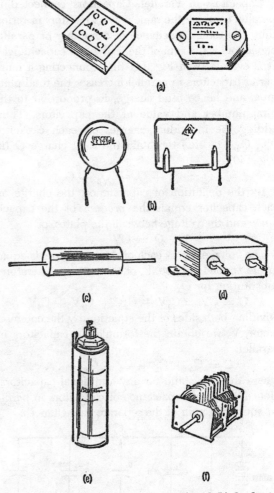

Fig. 140. Some Types of Capacitors—(a) and (b) fixed mica and ceramic capacitors; (c) and (d) paper by-pass capacitors; (e) electrolytic filter capacitor; (f) variable air-dielectric capacitor

by applying a d-c voltage between two electrodes of aluminum immersed in a chemical solution. The resulting electrolytic action forms a thin film of oxide on one of the electrodes, with a high dielectric constant and, hence, a large capacitance per unit area. However, since losses are high and increase with frequency and a **d-c polarizing voltage** is required, electrolytic capacitors are used only in low-frequency, d-c filter circuits, where a relatively large, inexpensive capacitance is required.

Finally, capacitors are used in **tuned or resonant circuits** to adjust the frequency. For these, air-dielectric capacitors are universally used, since air is an almost perfect dielectric (Fig. 140). The capacitor is made variable by meshing a set of rotable plates between another set of fixed stator plates.

Capacitors in Parallel. Capacitors connected in parallel combine the same way as resistors in series; thus, the total capacitance of a number of parallel capacitors is the sum of the individual capacitances. It is evident from Fig. 141 that connecting a number of capacitors in parallel increases the total plate area, and hence total charge, in proportion to the total number and value of the capacitors. Thus, adding the individual charges on each capacitor (Q_1, Q_2, Q_3, etc.) to obtain the total charge Q (in Fig. 140),

$$Q = Q_1 + Q_2 + Q_3 + Q_4$$

By the definition of capacitance, the charge on each capacitor equals the *product* of the capacitance and the voltage between the plates, or

$$Q = CV$$

The voltage across each of the capacitors is equal to the applied voltage, V, of the battery. Therefore, substituting for Q,

$$Q = CV = C_1V + C_2V + C_3V + C_4V$$

Dividing both sides of the equation by the common factor, V, we obtain the formula for capacitors in parallel:

$$C = C_1 + C_2 + C_3 + C_4 + \cdots$$

where the dots stand for any additional capacitors. Hence, the total capacitance of capacitors in *parallel* equals the *sum* of the separate capacitances.

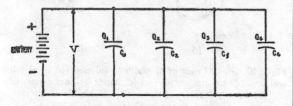

$$C = C_1 + C_2 + C_3 + C_4$$

Fig. 141. Capacitors in Parallel

EXAMPLE: If the capacitors in Fig. 140 above have the following values
$C_1 = 2$ μf; $C_2 = 20{,}000$ $\mu\mu$f; $C_3 = 0.005$ μf; and $C_4 = 0.25$ μf
what is the total capacitance of the parallel combination and what is the total charge stored, if the battery voltage is 60 volts?
Solution: $C = 2$ μf $+ 0.02$ μf $+ 0.005$ μf $+ 0.25$ μf $= 2.275$ μf. The total charge $Q = CV = 2.275 \times 10^{-6} \times 60 = 1.365 \times 10^{-4}$ *coulombs*.

Capacitors in Series. Capacitors in series combine in the same way as resistors in parallel. To see why this is so, consider a combination of four capacitors connected in series, as shown in Fig. 142. Since the current in a series circuit is everywhere

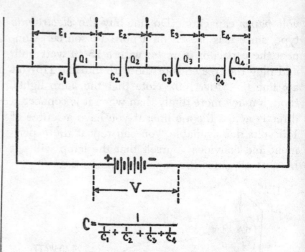

$$C = \frac{1}{\frac{1}{C_1} + \frac{1}{C_2} + \frac{1}{C_3} + \frac{1}{C_4}}$$

Fig. 142. Capacitors in Series

the same, the total charge ($Q = I\,t$) is equal to the charge on each of the capacitors, or

$$Q = Q_1 = Q_2 = Q_3 = Q_4$$

The sum of the voltage drops around a series circuit equals the applied voltage and, hence, the sum of the voltage drops across the capacitors equals the battery voltage, V. Thus,

$$V = E_1 + E_2 + E_3 + E_4$$

Let us divide both sides of this equation by Q, and we obtain

$$\frac{V}{Q} = \frac{E_1 + E_2 + E_3 + E_4}{Q} = \frac{E_1}{Q} + \frac{E_2}{Q} + \frac{E_3}{Q} + \frac{E_4}{Q}$$

But, by the definition of capacitance ($C = Q/V$),

$$\frac{V}{Q} = \frac{1}{C} \text{ ; moreover, } \frac{E_1}{Q} = \frac{1}{C_1} \text{; } \frac{E_2}{Q} = \frac{1}{C_2} \text{;}$$

$$\frac{E_3}{Q} = \frac{1}{C_3} \text{, and } \frac{E_4}{Q} = \frac{1}{C_4}$$

Substituting these relations in the equation above, we obtain

$$\frac{1}{C} = \frac{1}{C_1} + \frac{1}{C_2} + \frac{1}{C_3} + \frac{1}{C_4}$$

Thus, for capacitors in *series*, the *reciprocal* of the total capacitance is equal to the *sum of the reciprocals* of the individual capacitances. Moreover, solving for C:

$$C = \frac{1}{\frac{1}{C_1} + \frac{1}{C_2} + \frac{1}{C_3} + \frac{1}{C_4} + \cdots}$$

where the dots stand for any additional series capacitors. Therefore, corresponding to resistors in parallel, the total capacitance of a number of series-connected capacitors equals the **reciprocal of the** sum of the reciprocals of the individual capacitances. If all the capacitors are the same, the total capacitance is simply the value of one capacitor divided by the number of series-connected capacitors. Moreover, for *two* capacitors connected in series, we obtain the simple formula:

$$C = \frac{C_1 \times C_2}{C_1 + C_2}$$

It might strike you odd that anyone would want to connect capacitors in series, since the total capacitance obtained in this way is less than that of any individual capacitor. The reason is that series-connected capacitors *split the total applied voltage between them*, as we have seen (Fig. 142). Since the price of a capacitor goes up steeply with its voltage rating, it is often more economical in high-voltage circuits to series-connect a number of large capacitors of relatively low voltage rating rather than a single, lower capacitance of the required high voltage rating.

EXAMPLE 1: A 4-μf and a 12-μf capacitor are connected in series. What is the total capacitance of the combination?

Solution:

$$C = \frac{C_1 \times C_2}{C_1 + C_2} = \frac{4 \times 12}{4 + 12} = \frac{48}{16} = 3 \text{ microfarads.}$$

EXAMPLE 2: If the capacitors in Fig. 142 have the same values as those in the example for parallel capacitors (Fig. 141), what is the total capacitance of the series combination?

Solution: $\frac{1}{C} = \frac{1}{C_1} + \frac{1}{C_2} + \frac{1}{C_3} + \frac{1}{C_4}$

$$\frac{1}{C} = \frac{1}{2} + \frac{1}{.02} + \frac{1}{.005} + \frac{1}{.25} = 254.5$$

Hence, $C = \dfrac{1}{254.5} = 0.00393 \ \mu f = 3{,}930 \ \mu\mu f.$

Thus, the capacitance of the series combination is less than that of the smallest capacitor ($C_3 = 0.005$ μf).

Capacitive Reactance. A capacitor offers a certain opposition to the flow of alternating current, as we have observed in Experiment 24. In the case of a capacitor, this opposition is termed **capacitive reactance** (symbol X_C) and it is also measured in ohms. Let us see just what the magnitude of this capacitive reactance is.

We have seen that the total charge accumulated over a period of time, when a capacitor is connected to *d.c.*, is $Q = i \, t$, or equivalently, the current is the *rate of flow of charge* per unit time, $i = \frac{Q}{t}$. When a capacitor is connected to *a.c.*, the rate of flow of charge and, hence, the current is constantly *changing*. We therefore must take the ratio of a very small change in charge over a short period of time to obtain the instantaneous current at any time. This is written mathematically, $i = \frac{\Delta Q}{\Delta t}$, where Δ stands for "a small change." Since the charge is the product of the capacitance and the voltage impressed between the plates of the capacitor ($Q = CV$) we can substitute in the expression for the instantaneous current $\quad i = \dfrac{\Delta Q}{\Delta t} = \dfrac{\Delta CV}{\Delta t} = C\,\dfrac{\Delta V}{\Delta t}$

where we have taken C out of the expression, since it is a constant. This expression, then, tells us that to obtain the instantaneous value of the current we must multiply the capacitance by the *rate of change of the applied voltage with time*. Since the applied voltage is a *sine-wave a-c voltage* of the form

$$V = V_m \sin wt$$

we can easily compute its rate of change with time by the methods of elementary calculus. It is shown there that the rate of change (or *derivative*) of the above expression for V

$$\frac{dV}{dt} = w \, V_m \cos wt$$

where $\dfrac{dV}{dt}$ stands for the rate of change in calculus symbols. Equating this expression with the one previously obtained for the instantaneous current:

$$i = C \, \frac{\Delta V}{\Delta t} = wC \, V_m \cos wt$$

This resulting expression tells us that the instantaneous current in a capacitive a-c circuit varies as a *cosine wave;* that is, it has the same waveshape as the sine-wave voltage, but it *leads the voltage by an angle of 90 degrees*, since a cosine wave leads a sine wave by 90°. Moreover, the maximum (peak) value of the current, I_m, is obtained when cos $wt = 1$, so that we can write

$$I_m = wC \, V_m \text{ (substituting cos } wt = 1)$$

or $\dfrac{V_m}{I_m} = \dfrac{1}{wC}$

Finally, since the **ratio of the maximum values of the voltage and current equals the ratio of the ef-**

fective (rms) values (i.e.,

$$\frac{V_m}{I_m} = \frac{1.414\,V}{1.414\,I} = \frac{V}{I}),$$

we get the result

$$\frac{V}{I} = \frac{1}{wC}$$

Just as in the case of inductive reactance the ratio of voltage to current represents the opposition to the current, this ratio in a capacitive circuit defines its opposition to current flow or the *capacitive reactance*, X_C. Thus, we obtain the final result

$$\frac{V}{I} = \frac{1}{wC} = X_C$$

or $X_C = \dfrac{1}{wC} = \dfrac{1}{2\,\pi\,fC} = \dfrac{1}{6.283\,fC}$

where we have substituted for the angular velocity $w = 2\,\pi\,f$, as before. This expression shows that the **capacitive reactance of a circuit decreases with increasing capacitance and increasing frequency of the supply voltage.**

EXAMPLE 1: What is the capacitive reactance of a 0.002 μf capacitor at a frequency of 2.5 megacycles (2,500,000 cps)?

Solution: $X_C = \dfrac{1}{6.283\,fC} =$

$$\dfrac{1}{6.283 \times 2.5 \times 10^6 \times 0.002 \times 10^{-6}}$$
$= 31.8\ ohms.$

The example shows that when the capacitance is given in microfarads and the frequency in megacycles, the factors of 10^6 and 10^{-6} can be omitted, since they cancel out, and the result is obtained directly in ohms.

EXAMPLE 2: What is the magnitude of the current, when a 220-volt, 60-cycle a-c voltage is applied across a 25-μf capacitor?

Solution: The capacitive reactance

$$X_C = \dfrac{1}{6.28 \times 60 \times 25 \times 10^{-6}} = 106\ ohms.$$

Hence, $I = \dfrac{V}{X_C} = \dfrac{220}{106} = 2.075$ (rms) amperes.

Phase Relations. We already know that the current in a pure capacitance leads the impressed voltage by 90 electrical degrees or one-quarter cycle. Fig. 143 illustrates these phase relations in graphical form. The instantaneous waveforms show clearly that the current consists of a *cosine wave*, which

reaches its maximum points 90° earlier than the sine-wave voltage. This is also evident from physical considerations. When the voltage is first applied, the uncharged capacitor immediately draws a large charging current. But as soon as the potential between the plates of the capacitor reaches the value of the impressed voltage, the current drops to zero, since a capacitor cannot be charged to a voltage higher than that applied. In other words, the current is greatest at the beginning of the voltage cycle and becomes zero at the maximum value of the voltage. When the applied voltage starts to decrease from its peak value, the capacitor begins to discharge and the current flows in the *opposite* direction. This implies that the current *leads* the impressed a-c voltage by a 90°-phase angle.

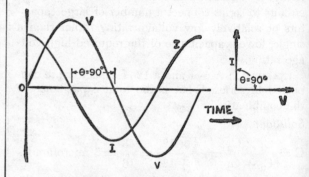

Fig. 143. Phase Relations in Pure Capacitance

The vector diagram at right of Fig. 143 summarizes these phase relations in simple form. Here the voltage (V) is the horizontal reference vector and the current (I) is drawn vertically upward, thus forming a *positive* or *leading* 90°-angle (ccw) with the voltage.

Power in Capacitive Circuit. Fig. 144 illustrates the **power graph** resulting when the instantaneous values of the voltage and current waves in a pure capacitance are multiplied by each other at various points along the time axis. The curve (P) has two positive and two negative loops during one complete cycle of the a-c voltage (V), exactly as in the case of a pure inductance (Fig. 138). Hence, we have again the situation that **no net power** is consumed, power being supplied to the capacitor and stored in its electric field during one-quarter (charging) cycle and the same amount being returned during the next quarter-cycle of capacitor discharge. But in any actual capacitor, a tiny amount of power is consumed because of the leakage of charge between the plates, which is the equivalent of a high resistance shunting the capacitor.

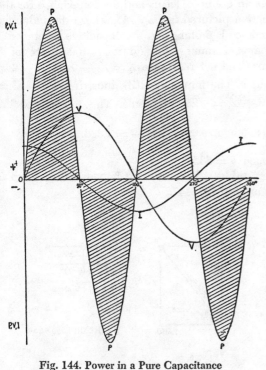

Fig. 144. Power in a Pure Capacitance

IMPEDANCE

We have seen that the alternating current in a circuit containing only *resistance* is *in phase* with the applied a-c voltage, while the *current in a pure inductance lags the impressed voltage by 90°* and that *in a pure capacitance leads the applied voltage by 90°*. What happens when an alternating voltage is applied to a circuit containing a combination of resistance and inductance, resistance and capacitance, or all three? We may venture a guess that the resulting alternating current would adjust itself to some value and assume a phase angle with respect to the voltage intermediate between the extremes (±90°), depending on the amounts of resistance, inductance and capacitance in the circuit. This guess is correct, but we have yet to find an effective method of calculating the magnitude and phase angle of the current in such a combination circuit. The concept of **impedance** (symbol Z), or **total opposition to the flow of alternating current in a circuit containing resistance, inductance, and capacitance,** has been devised to deal with this situation.

Impedance, or the total opposition to the flow of a.c., is a **vector quantity,** since it is composed of resistance—*in phase* with the emf of the source; and of reactance—*out of phase* with the applied emf. Because of the phase angle between resistance and reactance, the two *cannot be added arithmetically, but must be combined vectorially* to obtain the impedance. Moreover, the fundamentally different nature of resistance and reactance makes it impossible to add them directly. As we have seen, only the **resistance absorbs electric enegry** (and converts it into heat), while either inductive or capacitive **reactance store energy temporarily,** in the form of magnetic or electric fields, respectively. When resistance and reactance are added *vectorially,* the formula for the magnitude of the impedance (Z) of an a-c circuit becomes

$$Z = \sqrt{R^2 + X^2}$$

where R is the resistance and X is the *net reactance* of the circuit. According to this formula, the **impedance of an a-c circuit is equal to the square root of the sum of the squares of the resistance and the net reactance.** The impedance also has a *phase angle* as we shall see presently.

Net Voltage and Net Reactance. We have not yet explained the term **net reactance** (symbol X). Net reactance is simply the **vector sum** of all reactances or the **algebraic addition** of inductive and capacitive reactance (with regard to sign). Since the voltage across a pure inductance *leads* the current by 90° it is represented by a vector drawn *perpendicularly upward* from the start of the horizontal reference vector, as shown in Fig. 145. This upward vector, V_{x_L}, is considered *positive*, since it forms a *positive* (counterclockwise) angle with the reference vector. (Recall that the counterclockwise direction is taken as positive.) Similarly, since the voltage across a pure capacitance *lags* the current by 90°, the capacitive voltage is represented by a vector drawn *perpendicularly downward*, V_{x_C}. This vector is considered *negative* because it forms a *clockwise* angle of 90° with the reference vector. Since V_{x_L} and V_{x_C} are, thus, 180° out of phase with each other, they are *in line*, but in *opposite directions*. Their vector sum, which is the *net* voltage $(V_{x_L} - V_{x_C})$, may be found, therefore, either by *algebraic* addition $[V_{x_L} + (-V_{x_C})]$, or simply by subtracting V_{x_C} from V_{x_L}.

As shown in (*b*) of Fig. 145, the **net reactance** is found in exactly the same way as the net voltage. **Inductive reactance,** X_L, is considered *positive*, while *capacitive reactance,* X_C, is *negative*. The net reactance, X, thus is either the *algebraic sum* of the vectors $[X_L + (-X_C)]$, or simply the *arithmetic difference* of their magnitudes, $X_L - X_C$. In Fig. 145 $X_L = 6$ ohms and $X_C = 4$ ohms; hence the net reactance $X = X_L - X_C = 6 - 4 = 2$ ohms. We may

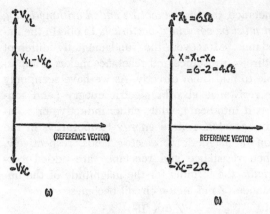

Fig. 145. Representation of Net Voltage (a) and Net Reactance (b)

substitute this result in the formula for impedance, obtaining

$$Z = \sqrt{R^2 + X^2} = \sqrt{R^2 + (X_L - X_C)^2}$$

If either inductance or capacitance are absent, X_L or X_C, respectively, drops out of the formula.

Impedance Triangle and Phase Angle. The form of the impedance equation indicates that the impedance vector may be represented as the *hypotenuse* (resultant) of a *right triangle*, the short sides of which are formed by the resistance and the reactance. Fig. 146 shows such an **impedance triangle**, with the resistance laid off along the horizontal side and the net reactance along the vertical side. As we have seen, the tangent of the acute angle (θ) is given by the ratio of the opposite side to the adjacent side of the triangle. Hence, the **tangent of the phase angle, θ** (in Fig. 146), is the **ratio of the net reactance to the resistance**, or

$$\tan \theta = \frac{X}{R} = \frac{X_L - X_C}{R}$$

The phase angle θ, thus, is the arctangent of this ratio, or

$$\theta = \arctan \frac{X}{R} = \arctan \frac{X_L - X_C}{R}$$

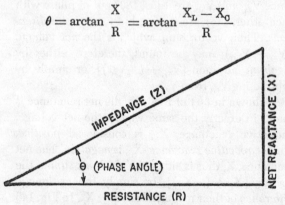

Fig. 146. The Impedance Triangle

As an example, let us find the impedance for the situation pictured in Fig. 145, where the inductive reactance is 6 ohms and the capacitive reactance is 2 ohms, assuming in addition a resistance of 3 ohms. The net reactance $X = X_L - X_C = 6 - 2 = 4$ ohms. The impedance (Z), thus, is $\sqrt{R^2 + X^2} = \sqrt{3^2 + 4^2} = \sqrt{25} = 5$ ohms. The tangent of the phase angle $\theta = \dfrac{X}{R} = \dfrac{4}{3} = 1.333$.

Hence $\theta = 53.2°$
(from tables). The resulting impedance diagram is pictured in Fig. 147.

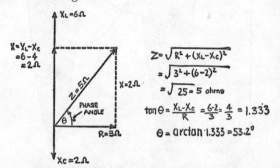

Fig. 147. Impedance Diagram for R = 3 ohms, X_L = 6 ohms, and X_C = 2 ohms

Ohm's Law for A.C. A modified form of Ohm's Law applies to alternating-current circuits, with the *resistance being replaced by the impedance*. Thus, we can state in a manner similar to Ohm's Law, that for an a-c circuit:

$$\text{current } I = \frac{E}{Z} = \frac{E}{\sqrt{R^2 + (X_L - X_C)^2}}$$

Further, the a-c voltage

$$E = IZ = I\sqrt{R^2 + (X_L - X_C)^2}$$

and the impedance $Z = \dfrac{E}{I}$

Moreover, the phase angle θ between the applied voltage (E) and the voltage drop in the resistance of the circuit (V_R) is the same as that in the impedance triangle; that is

$$\tan \theta = \frac{X}{R} = \frac{X_L - X_C}{R} = \frac{V_L - V_C}{V_R}$$

where V_L and V_C are the voltage drops in the inductance and capacitance, respectively. When either the impedance and resistance or the applied voltage (E) and the voltage drop across the resistance (V_R) are known, it is more convenient to obtain the phase angle θ from the relation for the *cosine* of a right triangle, which is defined as the *ratio* of the side adjacent to the angle θ to the hypotenuse. Hence, for the impedance triangle or a voltage triangle

$$\cos \theta = \frac{R}{Z} = \frac{V_R}{E}$$

This formula is an alternative way for finding the phase angle θ.

EXAMPLE: A series a-c circuit has a resistance of 90 ohms, an inductive reactance of 200 ohms and a capacitive reactance of 80 ohms. Voltmeters placed across the components reads 180 volts for the voltage drop across the resistance, 400 volts for the voltage drop across the inductance, and 160 volts for that across the capacitance. Find the impedance of the circuit, the applied emf (E), the phase angle, and the line current in the circuit.

Solution: Let us first construct the impedance triangle, illustrated in Fig. 148a. Since R = 90 ohms, $X_L = 20$ ohms and $X_C = 80$ ohms, the impedance $Z = \sqrt{R^2 + (X_L - X_C)^2} = \sqrt{90^2 + (200 - 80)^2} = \sqrt{22,500} = 150$ ohms. The tangent of the phase

angle: $\tan \theta = \frac{X}{R} = \frac{200 - 80}{90} = \frac{120}{90} = 1.333$.

Hence, from trigonometric tables, $\theta = \arctan 1.333$

$= 53.2°$. Equivalently, $\cos \theta = \frac{R}{Z} = \frac{90}{150} = 0.6$.

Again, from trigonometric tables,

$$\theta = \text{arc } \cos 0.6 = 53.2°.$$

V_C, perpendicularly downward. From the vector diagram it is evident that the applied emf, $E = \sqrt{V_R^2 + (V_L - V_C)^2} = \sqrt{180^2 + (400 - 160)^2} = \sqrt{90,000} = 300$ volts. As a check, $\cos \theta = \frac{V_R}{E} = \frac{180}{300} = 0.6$, and hence, $\theta = 53.2°$, as before. Finally,

the line current, $I = \frac{E}{Z} = \frac{300 \text{ volts}}{150 \text{ ohms}} = 2$ amperes.

Practice Exercise No. 12

1. What is the phase relationship between an alternating current flowing through a resistance and the voltage applied across it?

2. A 200-volt, 60-cycle a-c generator is connected across a resistive load of 25 ohms. What is the load current, the phase angle, and the true power consumed in the load?

3. What is the "skin effect"? What factors does it depend upon?

4. Coils of 250-mh, 350-mh, and 400-mh inductance are connected in series, spaced far apart. What is the total inductance? If the coils are then connected in parallel, what is their total inductance?

5. A 10-henry and a 20-henry choke are placed together so that their mutual inductance is 5 henrys. What is the total inductance if the coils are connected together (a) in series-aiding and (b) in series-opposing?

6. Explain the phase relationships between the applied emf, the counter emf, and the (alternating) current in an inductance.

7. What is inductive reactance and what does it depend upon? State the formula.

8. A 2-henry inductance coil is connected across a 100-volt, 50-cycle a-c supply. What is the inductive reactance? Neglecting coil resistance, what is the alternating current through the coil?

9. An inductance coil is connected in turn across two generators, which have a terminal voltage of 100 volts each. When connected across generator A the coil draws 2 amps; when connected across generator B it draws 50 amps. How do you explain this? Which of the two generators might be an a-c machine, which d.c.? What is the impedance of the coil; its d-c resistance and its inductive reactance? What is the power consumed in each case?

10. Why is a capacitor not a barrier to alternating current?

11. What is capacitive reactance? How does the effect of an increase in frequency on capacitive reactance differ from that on inductive reactance?

12. A 4-μf and a 6-μf capacitor are connected first in series and then in parallel. What is the capacitance in each case?

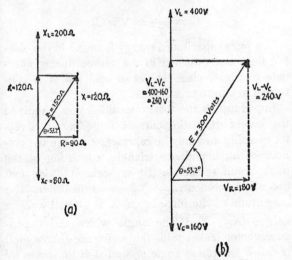

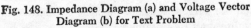

Fig. 148. Impedance Diagram (a) and Voltage Vector Diagram (b) for Text Problem

Now let us construct the voltage triangle (Fig. 148b), by laying off the resistive voltage drop, V_R, as horizontal reference vector, the inductive drop, V_L, perpendicularly up, and the capacitive drop,

13. A 100-$\mu\mu$f capacitor is connected across a 5-megacycle oscillator with a terminal voltage of 20 volts. What is the capacitive reactance and the current drawn from the oscillator?

14. What is the phase angle between the current and the impressed voltage in a capacitor? Is the current leading or lagging? How much power is consumed?

15. What is impedance? State the impedance formula.

16. The total resistance in a circuit is 50 ohms, the inductive reactance is 200 ohms, and the capacitive reactance is 80 ohms. What is the *net* reactance, the impedance, and the phase angle between impedance and resistance? Is the circuit inductive or capacitive?

SUMMARY

At high a-c frequencies current tends to flow near the surface of a conductor, thus increasing its resistance. The increase in the a-c resistance of a conductor, called the **skin effect**, is proportional to the diameter of the conductor and to the square root of the frequency.

Alternating current through a **resistance** is in phase with the applied voltage. Thus, Ohm's Law holds and $I = \dfrac{E}{R}$. The power consumed in the resistance, $P = EI = I^2R = \dfrac{E^2}{R}$.

The total inductance of a number of **inductors** connected **in series**, but *not* mutually coupled, equals the sum of the individual inductances. ($L = L_1 + L_2 + L_3 + L_4 + \ldots$) The inductance of two series-connected, mutually coupled coils is $L = L_1 + L_2 \pm 2M$, where M is the mutual inductance in henrys, the plus ($+$) sign is used for a series-aiding arrangement, and the minus ($-$) sign for a series-opposing connection.

The total inductance of coils in **parallel** equals the reciprocal of the sum of the reciprocals of the separate inductances. ($1/L = 1/L_1 + 1/L_2 + 1/L_3 + 1/L_4 + \ldots$)

The **counter emf** in an inductance coil is 180° out of phase with or opposed to the applied voltage. The alternating **current** through an inductance **lags** the applied emf by 90° in phase and **leads** the counter emf by 90°.

Inductive reactance is the opposition of an inductance to alternating-current flow; inductive reactance is **proportional** to the frequency and the inductance. ($X_L = 2\pi f L$.)

The (a-c) current through a pure inductance equals the applied voltage divided by the inductive reactance. $\left(I_L = \dfrac{V}{X_L} = \dfrac{V}{2\pi f L}.\right)$

In a pure inductance energy is stored in the magnetic field and then returned to the source during alternate quarter-cycles. No net or real power is consumed. An actual inductance (coil) contains some resistance and, hence, absorbs some amount of power.

A **capacitor** is a barrier to direct current, but **not** to A.C.

Capacitors connected **in parallel** combine like resistors in series, the total capacitance being the sum of the separate capacitances.
$$(C = C_1 + C_2 + C_3 + C_4 + \ldots)$$

Capacitors connected **in series** combine like resistors in parallel, the total capacitance being the reciprocal of the sum of the reciprocals of the separate capacitances.
$$\left(C = \frac{1}{1/C_1 + 1/C_2 + 1/C_3 + 1/C_4 + \ldots}\right)$$

The opposition of a capacitor to the flow of alternating current is called **capacitive reactance** (X_C); capacitive reactance decreases with increasing frequency and increasing capacitance.
$$\left(X_C = \frac{1}{2\pi f C}\right)$$

The current through a (pure) capacitance **leads** the applied voltage 90° in phase; its magnitude,
$$I_C = \frac{V}{X_C} = V w C.$$

In a pure capacitance energy is stored in the electric field and returned to the source during alternate quarter-cycles. No net or real power is consumed in a pure capacitance.

Impedance is the total opposition of a circuit to the flow of alternating current. Impedance is a vector quantity and may be represented as the hypotenuse of an impedance triangle, consisting of the total circuit **resistance** (R) as horizontal side and the **net reactance** ($X_L - X_C$) as vertical side. The magnitude of the impedance, Z, is given by $|Z| = \sqrt{R^2 + (X_L - X_C)^2}$. The angle which the impedance vector makes with the resistance (horizontal) is called the **phase angle** (θ) and it is the angle by which the current leads or lags the applied voltage. Its value may be obtained from:
$$\tan\theta = \frac{X}{R} = \frac{X_L - X_C}{R}, \text{ or}$$
$$\cos\theta = \frac{R}{Z} = \frac{V_R}{E} \text{ (series circuit).}$$

CHAPTER THIRTEEN

ALTERNATING CURRENT CIRCUITS

A.C. POWER

In the previous chapter we have learned something about the strange behavior of capacitors and coils, when subjected to alternating current flow. We have become acquainted with the concepts of "reactance" and "impedance" and have seen how Ohm's Law must be modified for use with A.C. We are finally ready to apply these new concepts to operating alternating-current circuits of varying complexity. But before we delve into this fascinating subject, let us digress briefly to see what meaning we can give to the term "power" in an alternating-current circuit, where reactance is present in addition to resistance. You may want to review the chapter on "ELECTRIC POWER AND HEAT" (Chapter 8) at this point.

When an alternating current flows through a resistance, the power consumed in it is the product of the current and the voltage across the resistance ($P = E \times I$), just as for direct current. In contrast, a *pure* inductance or a *pure* capacitance in an a-c circuit does *not consume any real power*, as we have seen, because the current is 90 degrees out-of-phase with the voltage in these components. In a-c circuits where a combination of resistance, capacitance and inductance is present, the current assumes a phase angle intermediate between zero and 90° and we would expect some amount of power to be consumed. When the products of the instantaneous current and voltage values are plotted for such an intermediate phase angle, it is found that the resulting power graph has positive lobes that are *larger* than the negative lobes. The power consumed by the circuit (or expended by the source) is then equal to the *difference between the areas* of the positive and negative power lobes.

You will find it cumbersome to plot the power graph from the instantaneous current and voltage values and then obtain the power by finding the difference between the areas of the positive and negative power lobes. In practice, the same result can be obtained far easier by drawing a vector diagram of the *effective* (rms) values of the current and the applied emf, and then computing the *amount of current that is in phase with the emf* (or the voltage drop across the resistance). The product of

the applied emf and the in-phase component of the total current is then the power expended. Fig. 149 shows such a vector diagram of the applied emf (E) and the current (I), which is out-of-phase with the voltage by the phase angle (θ). It is immediately apparent from the diagram that the in-phase component of the current is the horizontal projection of I upon E, or I cos θ. (In the current triangle, the ratio of the adjacent side to the hypotenuse

equals $\dfrac{I\cos\theta}{I} = \cos\theta$.) Similarly, the out-of-phase

component of the current is its *vertical* projection, or I sin θ. Forming the product of the applied emf (E) and the in-phase component of the current (I cos θ), we see that the true power expended in an a-c circuit is

$$P = E\,I\,\cos\theta$$

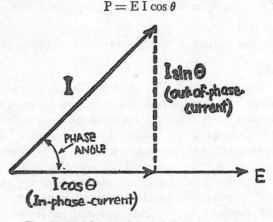

Fig. 149. In-Phase and Out-of-Phase Components of Current

EXAMPLE: Compute the power consumed by the circuit of the example (Fig. 148) in the last chapter.

Solution: We have already found that the applied emf, E = 300 volts, the line current, I = 2 amps., and cos θ = 0.6 for an angle of 53.2°. Substituting these values in the power formula:

P = EI cos θ = 300 × 2 × 0.6 = 360 watts power consumed.

Power Factor. Since the product of voltage and current must be multiplied by the cosine of the phase angle to obtain the power, cos θ is known as the **power factor** (abbreviated P.F.) of the circuit. Rewriting the relation for a-c power, we obtain for the power factor in an a-c circuit P.F. = cos θ =

$\frac{P}{EI}$; this is sometimes expressed as a percentage:

% P.F. $= \frac{P}{EI} \times 100$. Thus, we can determine the power factor of an a-c circuit by *dividing the watt-meter reading by the product of the voltmeter and ammeter readings*. The power factor is also given by the *ratio of resistance to impedance*

$$(\text{i.e., } \cos\theta = \frac{R}{Z}).$$

EXAMPLE: An a-c voltmeter across the line of an a-c circuit reads 220 volts and an ammeter in series with the line current reads 4 (rms) amps. If the wattmeter reading is 600 watts, what is the power factor of the circuit? What is its phase angle, impedance and resistance?

Solution: P.F. $= \frac{P}{EI} = \frac{600}{220 \times 4} = \frac{600}{880} = 0.682$ (Power Factor). Since $\cos\theta = 0.682$, the phase angle θ is found from tables to be 57°. The impedance $Z = \frac{E}{I} = \frac{220}{4} = 55$ ohms. Finally, since $\cos\theta = \frac{R}{Z}$, the resistance $R = Z \cos\theta = 55 \times 0.682 = 37.5$ ohms.

Reactive Power. The product EIcosθ is usually termed the true or real power of an a-c circuit to distinguish it from the apparent power obtained by multiplying the voltage by the current. The apparent power, expressed either in volt-amperes (va) or kilovolt-amperes (kva), contains *both* the true power expended in the resistance of the circuit, as well as the reactive power alternately stored and returned by the inductors and capacitors of the circuit. These power relations are conveniently portrayed by a **power triangle** (Fig. 150), consisting of apparent power (EI) as hypotenuse, real power as horizontal side, and reactive power as vertical side. From the simple trigonometry of a right triangle, we have the following:

Real Power = EI cosθ = Apparent Power × Power Factor

Reactive Power = EI sinθ = (Apparent Power) × sinθ

The reactive power (EIsinθ) of an a-c circuit is usually expressed in volt-amperes-reactive (vars) or in kilovolt-amperes-reactive (kvars).

EXAMPLE: The wattmeter reading in an a-c circuit with a power factor of 0.8 is 4.75 kw. What is the reactive power?

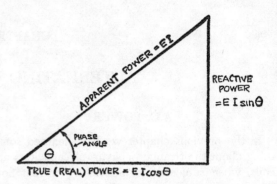

Fig. 150. A-C Power Triangle Showing Relations between True Power, Apparent Power and Reactive Power

Solution: The apparent power (EI) $= \frac{\text{Real Power}}{\text{P.F.}}$

$= \frac{4750}{0.8} = 5{,}940$ va. Since P.F. $= \cos\theta = 0.8$, the phase angle $\theta = 36.8°$; hence, the reactive power $=$ EI sin$\theta = 5{,}940$ sin 36.8° $= 5{,}940 \times 0.6 = 3{,}560$ vars.

We could have obtained the same result more simply by realizing that $\frac{\text{Reactive Power}}{\text{Real Power}} = \frac{\text{EI sin}\theta}{\text{EI cos}\theta}$ $= \tan\theta$, where $\theta = 36.8°$ in this case. Hence, Reactive Power = Real Power × tanθ = 4,750 tan 36.8° $= 4750 \times 0.75 = 3{,}560$ vars = 3.56 kvars.

Although the power companies charge their customers only for the *real* power consumed, they must make provisions for the additional *reactive* power that is alternately stored and returned by the electrostatic and magnetic fields associated with capacitors and inductors, respectively. The cables supplying the various circuits must be heavy enough to supply both the in-phase as well as the out-of-phase (reactive) currents circulating through them. The lower the power factor (cosθ), the greater is the apparent and reactive power demand and, hence, the heavier must be the wire gauge (copper) used. To reduce excessive apparent power demands and avoid waste of copper, the power companies try to keep the *power factor as close to unity as possible* (cos$\theta = 1$ for resistance). The power factor may be low because of a *lagging* phase angle due to highly inductive circuits, such as induction motors, chokes, transformers, etc. In these cases the power factor can be corrected by introducing large groups of capacitors in the circuit whose *leading* phase angle will cancel out the inductive lag. If the power factor is low because of a *leading* (capacitive) phase angle, as is the case for fluorescent lamps, for example, it may be corrected by introducing large inductors (called ballasts) into the circuit. Electrical machinery for a.c. is always rated in kilovolt-amperes, rather than watts, to permit estimating the current demands upon the associated circuits.

SERIES A-C CIRCUITS

Let us now apply the knowledge we have acquired about reactance, impedance and a-c power to some typical series a-c circuits. As is the case for d.c., the current in a series a-c circuits is everywhere the same. Moreover, the voltage drops across the various parts of the circuit, when added up *vectorially, equal the emf of the source.* (This is in contrast to d.c., where the *arithmetic* sum of the voltage drops equals the emf of the source.) For practical circuit calculations you should recall that the current flowing through an inductance *lags* 90° behind the applied voltage, while the current through a capacitance *leads* the applied voltage by 90°.

R-L Circuits. Consider first a simple series a-c circuit, consisting only of an inductance and a resistance, connected across an a-c generator (See Fig. 151.) This is actually the simplest possible inductive circuit, since any real inductor (choke coil) must have at least the resistance of its windings in series with the inductance.

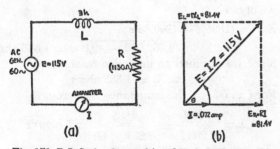

(a)   (b)

Fig. 151. R-L Series Circuit (a) and Vector Diagram (b)

Since there is no capacitive reactance, the impedance of this simple R-L circuit, $Z = \sqrt{R^2 + X_L^2}$,

and the current $I = \dfrac{E}{Z} = \dfrac{E}{\sqrt{R^2 + X_L^2}}$. Using the values given in the example of Fig. 151, where R = 1130 ohms, L = 3 henrys and E = 115 volts at 60 cycles, we obtain

$X_L = wL = 6.283 \, f \, L =$
$$6.283 \times 60 \times 3 = 1130 \text{ ohms}$$
$Z = \sqrt{(1130)^2 + (1130)^2} = \sqrt{2,556} = 1,600 \text{ ohms}$
and

$$I = \frac{E}{Z} = \frac{115}{1,600} = 0.072 \text{ ampere (approximately).}$$

The tangent of the phase angle (θ) between the applied voltage (E) and the current (I),

$$\tan\theta = \frac{X_L}{R} = \frac{wL}{R} = \frac{1130}{1130} = 1$$

Hence, from tables, the phase angle θ between E and I is 45°. We can also compute the voltage drop across the resistance,

$$E_R = I \, R = 0.072 \times 1130 = 81.4 \text{ volts}$$

and the voltage drop across the inductance,

$$E_L = I \, X_L = 0.072 \times 1130 = 81.4 \text{ volts}$$

Note that the *arithmetic sum* of the two voltage drops is 162.8 volts, which is greater than the applied emf, obviously an impossible situation. If we add the two voltage drops *vectorially*, however, as shown in the vector diagram of Fig. 151*b*, we obtain correctly

$$E = \sqrt{E_R^2 + E_L^2} = \sqrt{(81.4)^2 + (81.4)^2} =$$
$$\sqrt{13200} = 115 \text{ volts.}$$

To show the phase and power relations in the circuit of Fig. 151, we have plotted the instantaneous current (i), voltage (e) and their product—the instantaneous power (p)—in Fig. 152 below. Note that the current (i) in this circuit *lags* behind the instantaneous voltage (e) by a phase angle (θ) of about 45° or one-eighths cycle. Note further, that because of this phase angle, there are two small *negative* power lobes (each 1/8 cycle long) during each cycle of the impressed voltage, where power

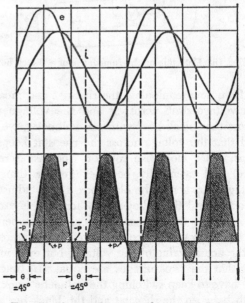

Fig. 152. Phase and Power Relations in R-L Series Circuit of Fig. 151

is being *returned* by the circuit to the supply, and there are also two large, *positive* power lobes (each 3/8 cycle long), where power is being *consumed* by the circuit or *expended* by the supply. Since the positive lobes are far greater in area than the negative ones, their difference is *positive* and, thus, real power is being consumed.

Rather than obtaining the average area of the positive power lobes, let us compute the power consumed by means of the power factor. The power factor of the circuit

$$PF = \cos\theta = \cos 45° = 0.707.$$

Hence, the real power consumed $= EI \cos\theta = 115 \times 0.072 \times .707 = 5.85$ watts. Equivalently, since all real power is consumed in the resistance,

$$P = E_R \times I = 81.4 \times 0.072 = 5.85 \text{ watts}$$

Further, the reactive power $= EI \sin\theta = 115 \times 0.072 \times .707 = 5.85$ vars. The apparent power is $E I = 115 \times 0.072 = 8.3$ volt-amperes.

As another example, let us make a paper experiment that will illustrate the power of the methods we have developed for calculating a-c circuits. Say we have a "black box" with two terminals sticking out, whose a-c characteristics we would like to determine using only a cheap voltmeter, an ordinary *d-c* ohmmeter, and a variable resistance (rheostat) of about 10,000 ohms. Let us connect the rheostat in series with the black box and the 120-volt, 60-cycle a-c power line, as illustrated in Fig. 153.

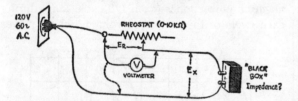

Fig. 153. Determining the Impedance of a "Black Box"

If we now manipulate the rheostat and measure the voltage drop across it and the voltage drop across the "black box" with our voltmeter, we will find that the voltage across the rheostat keeps increasing for clockwise rotation (increasing resistance) and eventually becomes greater than the voltage drop across the black box. Let us adjust the rheostat until the voltage drop across it is exactly the *same* as that across the box. An inexpensive a-c voltmeter will suffice, since we are not interested in the actual value of the voltage, but only in the fact that the two voltages are equal. (In practice, you have to keep switching the voltmeter back and forth between the rheostat and the black box after

each adjustment, until the point of equal voltage is reached.) As soon as the equal-voltage point is attained, disconnect the circuit and measure the d-c resistance (between the slider and fixed contact) of the rheostat with the ohmmeter. Also measure the d-c resistance between the terminals of the black box, if any.

Let us say, for the purposes of illustration, that the following values were obtained during a typical black box experiment:

Resistance of rheostat (for equal voltages)

$$R = 1,895 \text{ ohms}$$

d-c resistance (between terminals) of black box

$$R_x = 200 \text{ ohms}$$

Since the low d-c resistance of the black box indicates a continuous circuit for d.c., we shall assume that some form of inductance (coil) with a certain winding resistance is contained in the box. With the current in a series circuit everywhere being the same, and the voltage drops equal, we can write

$$E_R = E_X, \text{ and hence } I R = I Z_x;$$

where E_x and Z_x are the voltage drop and impedance, respectively, of the black box. Cancelling I, we have

$$R = Z_x = \sqrt{R_x{}^2 + X^2}$$

Squaring both sides:

$$R^2 = R_x{}^2 + X^2$$

and, hence,

$$X = \sqrt{R^2 - R_x{}^2}$$

Substituting the values obtained in the experiment, we obtain

$$X = \sqrt{(1895)^2 - (200)^2} =$$
$$\sqrt{3,593,000 - 40,000} = 1,885 \text{ ohms}$$

Since we assumed an inductive reactance,

$$X = X_L = 1,885 \text{ ohms}$$

For $f = 60$ cps, the inductance is, therefore,

$$L = \frac{X_L}{6.283 \text{ f}} = \frac{1885}{6.283 \times 60} = 5 \text{ henrys}$$

Our "black box" thus turns out to be the *equivalent* of a 5-henry choke coil in series with a 200-ohm resistance. It obviously does not matter whether or not we made a wrong assumption about an inductance being in the box. Even if the box contained a highly complicated network of inductors, capacitors and resistors connected in some series-parallel circuit, for all *practical purposes* at a frequency of 60 cycles it acts just like a 5-henry choke coil with 200 ohms winding resistance and, hence, *can be replaced by it.* The problem is not as remote as it might appear at first glance, since many actual inductors are "potted" in a shielded container (black box) with nothing but the terminals sticking out. If you wanted to determine the inductance of a transformer winding or of an unmarked choke coil, you can use the method we have just described.

R-C Circuits. Let us turn now to another simple series circuit containing only resistance and capacitance in series with an a-c generator. A typical R-C circuit with actual values is shown in Fig. 154a.

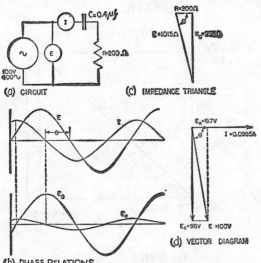

(a) CIRCUIT

(c) IMPEDANCE TRIANGLE

(b) PHASE RELATIONS

(d) VECTOR DIAGRAM

Fig. 154. R-C Series Circuit (a), Phase Relations (b), Impedance Triangle (c) and Vector Diagram (d)

Here a 100-volt, 400-cycle a-c generator is connected in series with a 0.4 microfarad capacitor and a 200-ohm resistor. First, we determine the capacitive reactance of the capacitor (C):

$$X_C = \frac{1}{2\pi fC} = \frac{1}{6.283 \times 400 \times 0.4 \times 10^{-6}}$$
$$= 995 \text{ ohms}$$

The impedance of the circuit is
$$Z = \sqrt{R^2 + X_C^2} = \sqrt{(200)^2 + (995)^2} =$$
$$\sqrt{40,000 + 990,000} = \sqrt{1,030,000} = 1,015 \text{ ohms}$$
Fig. 154c illustrates the impedance triangle constructed with a resistance of 200 ohms and a capacitive reactance of 995 ohms.

The cosine of the phase angle (θ), or power factor

$$\text{P.F.} = \cos\theta = \frac{R}{Z} = \frac{200}{1015} = 0.197$$

From tables, we find the phase angle $\theta = 78.63°$ Knowing the impedance, we can find the line current

$$I = \frac{E}{Z} = \frac{100}{1015} = 0.0985 \text{ ampere}$$

The voltage drop across the resistor,

$$E_R = IR = 0.0985 \times 200 = 19.7 \text{ volts}$$
and the voltage drop across the capacitor
$$E_C = IX_C = 0.0985 \times 995 = 98 \text{ volts}$$
When a vector diagram is constructed (Fig. 154d), using these voltage drops as the two sides, the resultant turns out to be the applied voltage of 100 volts, as expected.

Finally, the apparent power = E I = 100 × 0.0985 = 9.85 volt-amps. The real power consumed = E I $\cos\theta$ = 9.85 × 0.197 = 1.94 watts. The reactive power = E I $\sin\theta$ = 9.85 sin 78.63° = 9.85 × .98 = 9.65 vars.

The phase relations between the waveforms of the applied voltage (E) and the line current (I), and between the voltage across the capacitor (E_C) and that across the resistor (E_R) are illustrated in Fig. 154b.

It will be instructive to make another "black box" experiment, a capacitive one this time. The black box again is connected in series with a 0-10 kilohm rheostat and the 60-cycle a-c supply, exactly as the last time (See Fig. 153.) After adjusting the rheostat for equal voltage drops across it and across the terminals of the box, we obtain the following readings on the d-c ohmmeter:

Resistance of rheostat for equal voltages, R = 1328 ohms.

D-C resistance between terminals of black box, R_x = infinity (i.e., larger than could be measured with a simple ohmmeter).

Since the box presents a barrier to the flow of direct current, we shall assume this time that an essentially pure capacitance is contained in the box. Again we can write for a series circuit

$$E_R = E_x \text{ and, hence, } IR = IX.$$
(With an infinite d-c resistance, the box has reactance only.) Cancelling *I* on both sides, R = X, and since we assumed capacitive reactance,

$$R = X_C = \frac{1}{2\pi fC}$$

Hence, the capacitance of the box, $C = \dfrac{1}{2\pi fR}$

Substituting f = 60 cps, and R = 1328 ohms

$$C = \frac{1}{6.283 \times 60 \times 1328}$$
$$= 2 \times 10^{-6} = 2 \text{ microfarads.}$$

The box, thus, presents a capacitive reactance equal to that of a 2-μf capacitor. Again it does not matter whether a capacitor is *actually* contained in the box, or some combination of capacitance, inductance and (possibly) resistance. At the *single frequency* of 60 cycles the box acts just like a 2-microfarad capacitor and can be replaced by it.

R-L-C Circuits. A typical R-L-C circuit, such as that shown in Fig. 155a, combines the characteristics of the R-L and R-C circuits we have already discussed. Assume that a 100-volt 10,000-cycle (10-kc) a-c generator is connected to the R-L-C load shown in Fig. 155a. We would like to know the impedance into which the generator is "working." The impedance will tell us whether the circuit is primarily resistive, inductive or capacitive and we will then be able to determine the current and power requirements.

The total resistance of the series circuit is the sum of the individual resistors, R_1 and R_2:

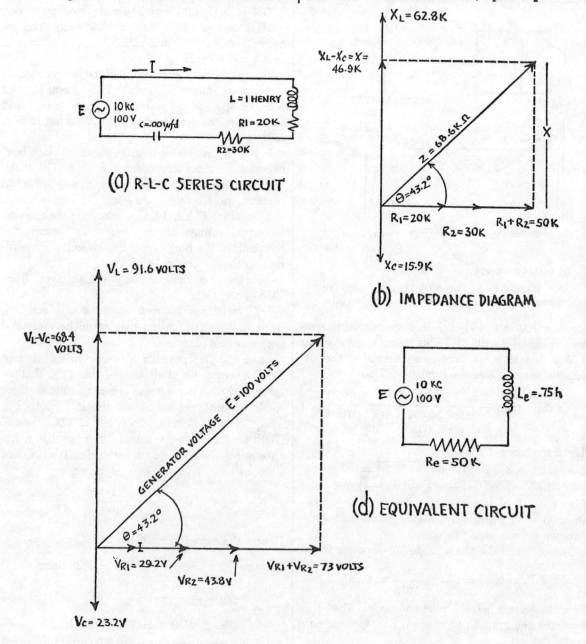

Fig. 155. R-L-C Series Circuit (a), Impedance Diagram (b), Voltage Diagram (c) and Equivalent Circuit (d)

$R = R_1 + R_2 = 20,000 + 30,000 =$
$$50,000 \text{ ohms} = 50 \text{ kilohms}$$
The inductive reactance of the 1-henry choke (L):
$X_L = 2\pi fL = 6.28 \times 10,000 \times 1 = 62,800$ ohms
The capacitive reactance of the 0.001 µfd capacitor

(C): $X_C = \dfrac{1}{2\pi fC} = \dfrac{1}{6.28 \times 10,000 \times .001 \times 10^{-6}}$
$$= 15,900 \text{ ohms}$$

The *net* reactance (X) is the difference between the inductive and capacitive reactances:
$X = X_L - X_C = 62,800 - 15,900 = 46,900$ ohms
Since the inductive reactance is greater than the capacitive reactance, the *net* reactance comes out *positive* and the circuit is primarily inductive. In the impedance diagram (Fig. 155b) we have laid off the resistance vectors along the horizontal (reference) direction and the inductive and capacitive reactance vectors along the vertical, in opposing directions. The impedance is the resultant (diagonal) of the parallelogram formed by the total resistance, $R_1 + R_2$, and the net reactance, $X_L - X_C$.

According to our formula, the *magnitude* of the impedance,
$Z = \sqrt{R^2 + (X_L - X_C)^2} = \sqrt{(50,000)^2 + (46,900)^2}$
$= \sqrt{25 \times 10^8 + 22 \times 10^8} = \sqrt{47 \times 10^8} =$
$$6.86 \times 10^4 = 68,600 \text{ ohms}$$
The tangent of the phase angle, θ:

$$\tan \theta = \frac{X}{R} = \frac{46,900}{50,000} = 0.938.$$

From tables we find the phase angle $\theta = 43.2°$. Equivalently, the *cosine* of the phase angle,

$$\cos \theta = \frac{R}{Z} = \frac{50,000}{68,600} = 0.729.$$

The phase angle θ again turns out to be 43.2° (from tables). If you measure the magnitude of the impedance vector (Z) in Fig. 155b and the phase angle θ, you will find that the graphical result checks the numerical computations. Thus we see that the generator works into an inductive load with an impedance of 68,600 ohms and a phase angle of 43.2°. As a matter of fact, we could replace the circuit of Fig. 155a at the generator frequency by an *equivalent* circuit that has but a single inductance, L_e, and a single resistor, R_e. (See Fig. 155d.) The value of the resistor is equal to the total resistance $(R_1 + R_2)$ of the circuit shown in (a); that is, $R_e = 50,000$ ohms. The equivalent inductance, L_e, must have the same reactance at 10 kc as the net reactance (X) of the original circuit shown in (a).

Hence: $2\pi fL_e = X$, or $L_e = \dfrac{X}{2\pi f} =$

$$\frac{46,900}{6.28 \times 10,000} = 0.75 \text{ hy.}$$

A 50,000-ohm resistor in series with 0.75-henry choke would therefore present exactly the same load to the 100-volt 10-kc generator as the original circuit of Fig. 155a.

Let us now find the line current (I) and the various voltage drops in the circuit. The current $I = \dfrac{E}{Z}$

$= \dfrac{100}{68,600} = 1.46 \times 10^{-3} = 1.46$ milliamperes. The current is *in phase* with the voltage across the resistors, or equivalently, is *out of phase* with the generator voltage by the amount of the phase angle. The voltage drop across resistor R_1,
$V_{R1} = I R_1 = 1.46 \times 10^{-3} \times 20,000 = 29.2$ volts
The voltage drop across resistor R_2,
$V_{R2} = I R_2 = 1.46 \times 10^{-3} \times 30,000 = 43.8$ volts
To construct the voltage diagram (Fig. 155c), we have laid off the two vectors, V_{R1} and V_{R2}, along the horizontal reference line, and also their vector sum, $V_{R1} + V_{R2} = 29.2 + 43.8 = 73$ volts. (The vector sum of vectors that are in line and in the same direction is the same as the arithmetic sum of their magnitudes.)

The voltage drop across inductor L, $V_L = I X_L = 1.46 \times 10^{-3} \times 62,800 = 91.6$ volts. Since this *voltage leads the current by 90°*, we have laid off vector V_L in the voltage diagram (c) perpendicularly upward, so that it forms an angle of $+90°$ with the current (I) and the resistive drops.

The voltage drop across capacitor, C, $V_C = I X_C = 1.46 \times 10^{-3} \times 15,900 = 23.2$ volts. Since the capacitive voltage drop *lags* behind the current by 90°, we have laid off vector V_C perpendicularly downward, so that it forms an angle of $-90°$ with the current and the resistive drops. The *net* reactive voltage drop in the circuit, $V_L - V_C = 91.6 - 23.2 = 68.4$ volts, and since this voltage is positive it is shown as a perpendicular upward vector in the voltage diagram.

Finally, the *vector* sum of the resistive and reactive voltage drops in the circuit must equal the applied generator voltage, or $E = \sqrt{(73)^2 + (68.4)^2} = \sqrt{5329 + 4671} = \sqrt{10,000} = 100$ volts. Checking the resultant vector diagram (c), you will find it to be 100 volts in magnitude (length) and, moreover, it *leads* the current (*I*) and the resistive voltage drops by the same phase angle we found previously; that is, $\theta = 43.2°$. Equivalently, the current in the circuit *lags behind* the generator voltage by a phase angle of 43.2°. To make this phase relation clear, the voltage vector diagram is frequently drawn with the generator voltage (*E*) along the horizontal reference line, which rotates the diagram clockwise by the amount of the phase angle. The current then forms a *negative* or *lagging* angle (—43.2° in this case) with the applied generator voltage, *E*.

In conclusion, let us explore the power relations in the circuit of Fig. 155a. We have already seen that the power factor, P.F. = $\cos\theta = 0.729$, which is rather low (compared to 1). The *apparent* power = E I = $100 \times 1.46 \times 10^{-3}$ = 0.146 volt-ampere
The *real* power = EI $\cos\theta$ = 0.146×0.729 = 0.1063 watt
The *reactive* power = EI $\sin\theta$ = 0.146 sin 43.2° = 0.146×0.685 = 0.1 var

Summarizing the most important results of our analysis, we see that the circuit of Fig. 155a offers an impedance of 68,600 ohms, draws a current of 1.46 ma that *lags* 43.2° behind the generator voltage, and consumes 0.1063 watt real power.

PARALLEL A-C CIRCUITS

We do not have to learn anything new to deal with parallel a-c circuits; we need only combine our knowledge of simplifying *d-c* parallel circuits, vectors, impedance, and power calculations. The important fact, you will recall, that is true for *any* parallel circuit is that the **potential difference across each branch of the circuit is the same and is equal to the emf of the source**. We need therefore only divide this common potential difference by the "opposition" (i.e., *impedance*) offered by each branch to determine each of the separate branch currents. Since impedance has a phase angle, each of the branch currents generally will have a phase angle with respect to the line voltage or the current in a resistive branch. Hence, to find the total "line" current supplying the parallel branches, we must compute the *vector sum* of the individual branch currents, rather than the *arithmetic* sum (as we did for parallel d-c circuits). Once we have the total line current (I) in the circuit, we can easily obtain the total impedance of the circuit by dividing the applied emf by this current ($Z = \dfrac{E}{I}$). The resulting total impedance, moreover, will permit us to replace

the actual parallel branches by an *equivalent simple series circuit*. The power calculations for a parallel circuit are the same as for a series circuit, once the phase angle between the applied emf and the total line current is known.

If the parallel a-c circuit consists only of *resistive* branches all the branch currents will be *in-phase* with the applied voltage, and the calculations will then be exactly the *same* as for a parallel d-c circuit with the same applied voltage and the same resistive branches. If the parallel circuit consists of a resistive and a capacitive branch, the current in the capacitive branch will *lead* the applied voltage by 90 degrees in phase, while that in the resistance will be *in-phase* with the applied voltage. As a consequence, the total current supplying the circuit will *lead* the applied voltage by a phase angle somewhere between zero and (plus) ninety degrees. The upshot is that the total circuit impedance is made up of resistance and *capacitive* reactance and, hence, the parallel circuit can be replaced (for the purpose of calculation) by an **equivalent series circuit**, consisting of a capacitor and resistor. Finally, if the parallel circuit consists of a resistive and an inductive branch, the current in the inductive branch will *lag behind* the applied voltage by 90°, while the current in the resistive branch will be *in-phase* with the applied voltage. The result is that the total current supplying the parallel circuit will *lag* behind the applied voltage by a phase angle somewhere between zero and (minus) ninety degrees. Moreover, the total circuit impedance is made up of resistance and **inductive** reactance, so that the parallel circuit can be replaced by an **equivalent** series circuit, consisting of an inductance (coil) and a resistance. The most interesting case is when resistance, capacitance and inductance are all present in the parallel circuit. It is not possible, then, to make a general prediction concerning the phase angle and the nature of the total circuit impedance.

R-L-C Parallel Circuit. Fig. 156 illustrates a typical parallel a-c circuit consisting of resistive, capacitive and inductive branches. Let us compute the branch currents, the total current, the total impedance and the power consumed for the circuit values illustrated.

First, we obtain the three branch impedances, Z_1, Z_2 and Z_3: $Z_1 = R = 6$ ohms; $Z_2 = X_C = \dfrac{1}{2\pi fC}$

$$= \frac{1 \times 10^6}{6.28 \times 100 \times 66} = 24 \text{ ohms;}$$

and $Z_3 = X_L = 2 \pi f L = 6.283 \times 100 \times 19 \times 10^{-3}$ = 12 ohms.

Substituting in Ohm's Law for A.C. $(I = \dfrac{E}{Z})$, we obtain for the individual branch currents, I_R, I_C and I_L:

$$I_R = \frac{48}{6} = 8 \text{ amps}; \ I_C = \frac{48}{24} = 2 \text{ amps}; \text{ and}$$

$$I_L = \frac{48}{12} = 4 \text{ amps}.$$

To obtain the total current, I_t, which is the *vector sum* of the branch currents, we construct the current vector diagram, illustrated in Fig. 156. Since the generator voltage (E) is the common voltage impressed across all the branches, we use it as the horizontal reference vector in our diagram. The current through the resistance, I_R, is *in phase* with the applied voltage and, hence, this current vector is laid out along the horizontal also. The current

through the capacitor *leads* the impressed voltage (E) by 90°; we therefore draw the current vector I_C perpendicularly upward along the vertical, from the same starting point as the voltage vector. Finally, the current flowing in the inductance *lags* the impressed voltage by 90° and, hence, we lay out current vector I_L perpendicularly downward from the starting point of the voltage vector.

It is evident that the current through the capacitor (I_C) is 180° out of phase with the current through the inductor (I_L); that is both are in line, but in opposite directions. The vector sum of these two reactive currents, therefore, is simply their arithmetic difference, $I_L - I_C$. Since I_L is greater than I_C, this difference is positive and ($I_L - I_C$) is an inductive or *lagging* vector. The final step is to obtain the vector sum of ($I_L - I_C$) and I_R, the resultant of these two vectors being the total current, I_t. As you can see from the diagram, this resultant is the diagonal of the parallelogram formed

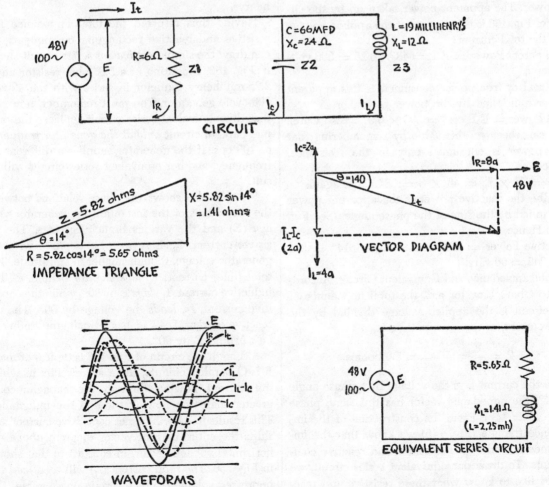

Fig. 156. R-L-C Parallel Circuit, Vector Diagram, Impedance Triangle, Waveforms and Equivalent Series Circuit

between vectors $(I_L - I_C)$ and I_R, or equivalently, it is the *hypotenuse of a right triangle*, of which the two sides are I_R and $(I_L - I_C)$. The length of this hypotenuse, you will recall, equals the square root of the sum of the squares of the two sides. Accordingly, the *magnitude* (length) of the total current vector,

$$I_t = \sqrt{I_R^2 + (I_L - I_C)^2} = \sqrt{(8)^2 + (4-2)^2} = \sqrt{64 + 4} = \sqrt{68} = 8.25 \text{ amps.}$$

Phase Angle and Power Factor. It is also evident from the current vector diagram (Fig. 156) that the cosine of the phase angle (θ) between the total current (I_t) and the impressed voltage (E) is simply the *ratio* of the adjacent side to the hypotenuse, or the resistive current (I_R) to the total current (I_t). This is also the power factor. Hence, the power factor

$$\text{P.F.} = \cos\theta = \frac{I_R}{I_t} = \frac{8}{8.25} = 0.97$$

From tables we obtain the phase angle
$$\theta = \text{arc } \cos 0.97 = 14°$$

Power. The apparent power taken up by the circuit of Fig. 156 is the product of the applied voltage and the total current:

Apparent Power $= E\, I_t = 48 \times 8.25 = 396$ volt-amperes

The real or true power consumed is this apparent power multiplied by the power factor, or

Real Power $= E\, I_t \cos\theta = 396 \times 0.97 = 384$ watts. You can check on this value by remembering that real power is consumed only in the resistance. Hence,

$$P = E \times I_R = 48 \times 8 = 384 \text{ watts again.}$$

Finally, the reactive power is the apparent power multiplied by the *sine* of the phase angle. (See Fig. 150.) Hence,

Reactive Power $= E\, I_t \sin\theta = 396 \sin 14° = 396 \times 0.242 = 96$ vars.

Total Impedance and Equivalent Circuit. According to Ohm's Law for a.c., the total impedance of the circuit is the applied voltage divided by the total current, or

$$Z = \frac{E}{I_t} = \frac{48}{8.25} = 5.82 \text{ ohms}$$

Since the current lags the voltage by a phase angle of 14°, the impedance vector has this same phase angle ($\theta = 14°$). From the construction of the impedance triangle (Fig. 146) we know that the impedance is made up of **resistive and reactive components**. To draw our **equivalent** series circuit, we would like to know what these resistive and reactive components are. This means that we have to *resolve* the impedance into two rectangular component vectors, as we have learned in Chapter 11.

Fig. 156 shows the impedance triangle obtained by resolving the impedance vector into two rectangular components. Here we have drawn the impedance vector with a *length* equal to its magnitude of 5.82 ohms and with a *direction* that is inclined with respect to the horizontal by the amount of the phase angle ($\theta = 14°$). The horizontal component represents the total resitance and it equals
$$R = Z \cos\theta = 5.82 \times 0.97 = 5.65 \text{ ohms}$$
The vertical component represents the total reactance and it equals
$$X = Z \sin\theta = 5.82 \times 0.242 = 1.41 \text{ ohms}$$
Since the total current *lags* the impressed voltage, we know that this reactance must be *inductive* and, hence, we can write
$$X_L = 1.41 = 2\pi f\, L = 628.3\, L$$

Hence, $L = \dfrac{1.41}{628.3} = 2.25 \times 10^{-3} = 2.25$ milli-henrys.

Having thus resolved the total impedance into resistive and reactive rectangular components, we can draw the simple *equivalent* series circuit shown in Fig. 156, consisting of a 5.65-ohm resistor and a 2.25-millihenry inductor in series with the 48-volt, 100-cycle generator. You must remember, however, that this equivalent series circuit replaces the original parallel circuit *only at the generator frequency* of 100 cps. If the generator supplies a different a-c frequency, another equivalent series circuit will result.

Fig. 156 also shows the phase relations between the waveforms of the instantaneous generator voltage (E) and the various branch currents. The resistive current, I_R, is seen to be *in-phase* with the generator voltage, though smaller in amplitude. The total (line) current, I_t, *lags slightly* behind E. The inductive current, I_L, *lags* E by 90°, while the capacitive current, I_C, *leads* the voltage by 90°. The *net* reactive current, $I_L - I_C$, has a small amplitude and *lags behind* E by 90°.

Although the circuit of Fig. 156 is typical of many R-L-C parallel circuits, there are circuits in which the current flowing through the capacitance is greater than that flowing through the inductance. This results in drawing a **net capacitive current** and shifts the entire current vector diagram *above* the horizontal voltage vector, in contrast to that shown in Fig. 156. The total current (I_t) will then *lead* the impressed voltage (E) by a certain phase angle. The formula for the *magnitude* of the total current

$(I_t = \sqrt{I_R{}^2 + (I_L - I_0)^2})$, however, is still good, since even with I_0 larger than I_L, the *square* of the negative quantity $(I_L - I_0)$ is still a *positive* number. The equivalent series circuit will then contain a capacitor instead of the inductor.

Practice Exercise No. 13

1. Explain the significance of the power factor. What is its range of values? Do power companies desire a high or a low P.F.?

2. Draw a power triangle and explain the relations between apparent power, real or true power, and reactive power.

3. A wattmeter in an a-c circuit reads 4000 watts, an a-c voltmeter across the line reads 120 volts and an ammeter inserted in series with the circuit reads 50 amperes total current. What is the apparent power, the power factor, the true power, and the reactive power?

4. A 0.1-henry coil with a winding resistance of 18 ohms is connected across a 220-volt, 60-cycle power source. What is (a) the reactance of the coil; (b) its impedance; (c) the current through the coil; and (d) the phase angle by which the current lags the applied voltage?

5. A "black box" is connected in series with a rheostat across the 60-cycle a-c supply. When the rheostat is adjusted for equal voltage drops across it and the black box, the d-c resistance of the rheostat measures 3770 ohms and that of the black box 25 ohms. What is the impedance of the black box and by what could it be replaced?

6. A capacitor of 80 ohms reactance is connected in series with a 50-ohm resistance across a 100-volt, 60-cycle a-c line. Calculate the current in the circuit, the phase angle by which the current leads the applied voltage, the capacitance, the real power consumed and the reactive power.

7. A series R-L-C circuit consisting of a 200-ohm resistor, a 20-millihenry choke coil, and a 0.36-μf capacitor is connected across a 20-volt, 1000-cycle a-c voltage. Compute the inductive reactance, the capacitive reactance, the net reactance, the impedance, the line current, the phase angle and state whether the current leads or lags the applied emf. Also compute the line current, when the capacitor is removed from the circuit and that when the choke coil is removed from the circuit.

8. A 100-ohm resistor, a 0.25-henry inductance coil with a winding resistance of 40 ohms and a 17.7-μf capacitor are connected in *series* across a 220-volt, 60-cycle a-c power line. Find the potential difference across the resistor, the inductance coil, and the capacitor. Also calculate the total impedance, the line current, the phase angle, the apparent power and the true power consumed.

9. The circuit of problem 8 is disconnected and the resistor, inductive coil and capacitor are connected *in parallel* across a 12-volt 60-cycle line. What is the current in each branch, the total line current, the total impedance, and the true power consumed?

SUMMARY

The product of voltage and current (volt-amperes) in an a-c circuit is called the apparent power (E I). To obtain the real or true power consumed by the circuit, the apparent power must be multiplied by the power factor, which equals the cosine of the phase angle between voltage and current. (P.F. = $\cos\theta$; $P = EI \cos\theta$.)

The power factor may also be determined by dividing the wattmeter reading by the product of the voltmeter and ammeter readings. (P.F. = $\cos\theta = \dfrac{P}{EI}$.) Power companies desire a *high* power factor.

Reactive power in an a-c circuit is the product of apparent power and the sine of the phase angle ($P_{reac} = EI \sin\theta$). It is measured in vars.

The power relations in an a-c circuit may be portrayed by a (right) power triangle, with apparent power as hypotenuse, real power as horizontal side and reactive power as vertical side.

Ohm's Law applies to A.C., when resistance is replaced by impedance. Thus, the current is the applied voltage divided the total impedance.

$$\left(I = \frac{E}{Z}; E = I Z, \text{ and } Z = \frac{E}{I} = \sqrt{R^2 + (X_L - X_0)^2}\right)$$

In an alternating-current series circuit, the total current $\left(I = \dfrac{E}{Z}\right)$ is everywhere the *same* and the vector sum of the voltage drops across the separate parts of the circuits equals the applied emf. The total impedance is the vector sum of the separate resistances and reactances in the circuit. The total (line) current leads or lags the applied emf, depending upon whether the net reactance $(X_L - X_0)$ is inductive (+) or capacitive (—).

In an alternating-current parallel circuit, the common potential difference applied across each branch of the circuit equals the emf of the source. The current in each branch is this common voltage divided by the impedance of the branch. The total current is the vector sum of the branch currents. The total impedance is the applied emf divided by the total (line) current $(Z = E/I_t)$.

CHAPTER FOURTEEN

RESONANCE

If you have ever taken a pretty girl on a leisurely boat ride on a balmy summer night, you may know what "resonance" is—a feeling of "being in tune," "having sympathetic vibrations" and pleasant harmony between you and the maiden. Although the romantic version of resonance defies an exact definition, the state of "being in tune" and having "sympathetic vibrations" describes pretty well the principle of physical resonance, which is a universal mechanical and electrical phenomenon.

Ordinary physical objects best illustrate the principle of resonance. Every object has a "natural" frequency of vibrations, which depends on its dimensions and on its mass. When you strike a key on a piano, the hammer "excites" a string of a certain length, mass and tension to its natural frequency of vibrations and it gives forth a tone of corresponding pitch. The piano tone may excite a nearby vase into vibrations, resulting in unpleasant "resonance." The vibrations of the vase occur only when the *frequency* of the piano tone is the *same as the natural frequency* of the vase, and the two are said to be *in tune* or *in resonance*. Moreover, as long as the same key is struck repeatedly, *energy will continue to be transferred to the vase* and its *resonant vibrations will be maintained*.

As another example, soldiers marching across a bridge in step and at a certain cadence may cause the bridge to *vibrate at its natural frequency*. If the constant small impulses from the marching feet take place at the same frequency as the natural frequency of bridge oscillations, *resonance takes place* and the bridge is *forced into oscillations at its natural frequency*. Since the effect is cumulative, the amplitude of vibrations may become so large that the bridge will be destroyed. This is the reason why soldiers are told to break step when crossing a bridge.

Electrical Oscillations and Resonance. A weight on a spring, when pulled downward and released, will oscillate freely as a pendulum at its **natural frequency**, which depends on the weight and the elasticity of the spring (See Fig. 157). If you pull the weight each time it approaches its bottom position, you can produce very large, *continuous* oscillations, which die down only after you stop the rhythmic impulses. Again, by adjusting the fre-

quency and the timing of your impulses to the natural frequency of the oscillator, you have produced *resonance* with consequent large-amplitude oscillations. You can produce *forced* oscillations of the spring at some other frequency by varying the rhythm of your pulling, but the *amplitude* (maximum displacement) of these forced oscillations will be much smaller than those produced at the natural frequency of oscillations.

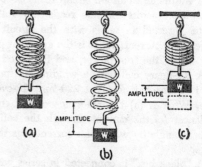

Fig. 157. Mechanical Spring Pendulum

It is shown in advanced texts (e.g., See **Electronics Made Simple**) that electrical circuits containing inductance and capacitance comprise **oscillating systems** similar to the mechanical spring pendulum. The **inductance** of a coil, which resists any change of the current, is analogous to the inertia of a mass or weight that resists any sudden change in its motion. Similarly, the charge on a **capacitor** is analogous to the displacement and tensioning of the spring. The capacitance and elasticity of the spring are also (inversely) related.

When you extend the spring of the mechanical pendulum by pulling the weight, you *store potential energy* in it in the form of *tension*. Analogously, placing a charge on a capacitor, *stores* potential (electric) *energy in the electric field* between the plates. When you let the weight go, the tension or potential energy is *released* into the (kinetic) *energy of motion of the weight*. Moreover, because of the mechanical inertia or flywheel effect of the weight, it does not stop when the spring is slack, but continues moving until the spring is compressed again and the energy of motion is once again stored as potential energy in the compressed spring. Similarly, if you discharge a capacitor through an inductance coil (in a closed circuit) the energy of the electric field is released by the motion of the charges (i.e. current) through the coil, which build up a **magnetic field** about the coil. The energy in the electric field of the capacitor is thus temporarily stored in the magnetic field of the coil. And because the inductance (inertia) of the coil resists any change in the current (motion of charges), the current does not stop when the capacitor is fully discharged, but continues to flow until the capacitor is *recharged in the opposite direction* and the energy is again stored in its electric field. Just as the mechanical pendulum continues to oscillate by the alternate storage and release of mechanical energy in the spring and weight, respectively, the electrical L-C circuit continues to oscillate by the alternate storage and release of energy in the fields of the capacitor and coil. The mechanical pendulum stops oscillating when all its energy is used up in the *friction* of the spring and bearings; similarly, an inductive-capacitive circuit stops oscillating when all its electrical energy is used up in the inevitable resistance present in the coil winding and conductors.

The analogy between mechanical and electrical circuits explains how an L-C circuit can function as an **oscillating system**, when d-c pulses of energy are fed to it with the right timing to excite its natural frequency of oscillations. In texts on electronics it is explained how such an electrical *oscillator* is capable of *converting direct to alternating current* at the frequency of natural oscillations. What happens when current that is already alternating is fed to such an L-C oscillating system? The situation is analogous to pulling the weight of the spring pendulum back and forth at a frequency that is *not* its natural frequency, thus *forcing* it to *oscillate at the impressed frequency*. This can be done, of course, but the amplitude of the oscillations (displacements) is strictly limited by the opposition of the spring and the weight. Similarly, forcing an electrical L-C circuit to oscillate at some arbitrarily imposed frequency, limits the amplitude of the oscillating (alternating) current to that permitted by the combined opposition (impedance) of the coil and the capacitor.

However, when the frequency of the force (voltage) impressed on the spring pendulum or the L-C circuit *equals the natural frequency* of either oscillating system, resonance is produced, and the amplitude of the oscillations is limited only by the (mechanical or electrical) resistance of the system. At resonance, only sufficient energy need be fed to such an oscillating system to overcome the losses

in the resistance in order to maintain the oscillations at their large, resonant amplitude. Thus, one of the significant effects of resonance is that it permits the **most efficient possible transfer of energy** from a power source to a load. Another and even more important result of resonance is the fact that a resonant circuit responds to or "selects" only a **particular (resonant) frequency and rejects all others.** This **selectivity** of resonant circuits makes them highly useful, in conjunction with electron tube amplifiers, for the selection and amplification of a specfic radio frequency, or a narrow band of frequencies. (See **Electronics Made Simple**)

SERIES RESONANCE

We have seen that the **inductive reactance (X_L)** of a coil *increases with frequency*, while the **capacitive reactance (X_C)** of a capacitor *decreases with frequency* (See Fig. 158). At some frequency, f_r, they must both be *equal* and the **net reactance** ($X_L - X_C$) will equal *zero*. At this **resonant frequency** f_r, therefore, the total opposition to current flow (impedance) of an R-L-C circuit is simply the *resistance*. Since the net reactance is zero, the current at resonance is *in phase* with the applied voltage ($\theta = 0$) and is given by Ohm's Law ($I = E/R$). Since the **impedance is equal to the resistance** at the resonant frequency, it is obviously at a *minimum* and, hence, the line *current supplying the circuit is a maximum.*

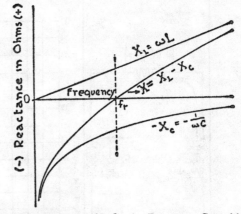

Fig. 158. Variation of Inductive Reactance, Capacitive Reactance and Net Reactance with Frequency

Conditions for Series Resonance. Let us summarize the conditions for resonance in an R-L-C series circuit in mathematical form. At resonance:
Reactances: $X_L = X_C$ (Inductive Reactance equals capacitive reactance)
Net Reactance: $X = X_L - X_C = 0$ (Net reactance is zero)

Impedance: $Z = \sqrt{R^2 + X^2} = \sqrt{R^2 + 0} = R$

(Impedance equals resistance and is *minimum*)

Total (line) Current: $I = \dfrac{E}{Z} = \dfrac{E}{R}$

(Current follows Ohm's Law and is *maximum*)

Power Factor: $P.F. = \cos\theta = \dfrac{R}{Z} = \dfrac{R}{R} = 1$

(Power Factor is *unity*)

Phase Angle: $\theta = \arccos 1 = 0°$

(Phase angle is zero; current is *in phase* with voltage)

Resonant Frequency. We can easily derive the resonant frequency, f_r, at which resonance occurs and these conditions prevail. By definition, at the resonant frequency, f_r

$$X_L = X_C$$

Hence, $2\pi f_r L = \dfrac{1}{2\pi f_r C}$ (Substituting for X_L and X_C)

$$f_r^2 = \dfrac{1}{4\pi^2 L C} \text{ (Cross-multiplying)}$$

and $f_r = \dfrac{1}{2\pi \sqrt{L C}}$ (taking the square root)

where f_r is in cycles per second, L in henrys and C in farads. By substituting for the constant $1/2\pi$, this may be written

$$f_r = \dfrac{159.1}{\sqrt{L C}}$$

where *either* f is in cps, L in henrys, and C in microfarads (μf) or f is in megacycles (mc), L in microhenrys (μh) and C in micromicrofarads ($\mu\mu f$).

EXAMPLE 1: A 2-henry inductance coil is connected in series with a 10-microfarad capacitor and both are placed across a variable-frequency a-c source. At what frequency will resonance occur?

Solution: $f_r = \dfrac{159.1}{\sqrt{L C}} = \dfrac{159.1}{\sqrt{2 \times 10}} = \dfrac{159.1}{\sqrt{20}} =$ 35.6 cps.

EXAMPLE 2: A 5-$\mu\mu f$ capacitor and a coil are placed in series in a circuit designed to resonate at 50 megacycles. What should the inductance of the coil be to attain this resonant frequency?

Solution: $f_r^2 = \dfrac{(159.1)^2}{L C} = \dfrac{25,330}{L C}$;

hence, $L = \dfrac{25,330}{f_r^2 C} = \dfrac{25,330}{(50)^2 \times 5} = \dfrac{25,330}{2500 \times 5}$ $= 2.02$ microhenrys.

Series-Resonant Circuit and Resonance Curve. A typical series-resonant circuit, consisting of resistance, inductance, and capacitance, is illustrated in Fig. 159. Here a 150 microhenry coil is connected in series with a 169-micro-microfarad capacitor and a resistor to a 1-volt, variable-frequency

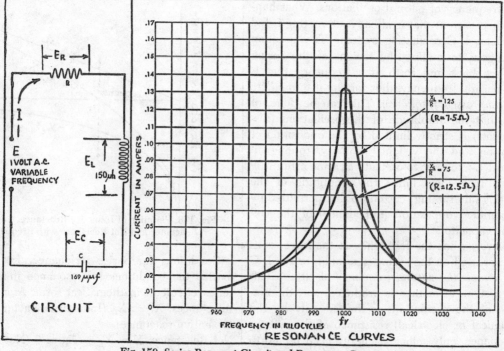

Fig. 159. Series-Resonant Circuit and Resonance Curves

a-c source. The resistance (R) represents the winding resistance of the coil, the "leakage" resistance of the capacitor, plus any additional resistance (such as that of the wires) in the circuit.

Let us first determine the resonant frequency of this circuit, using the simplified formula for L in microhenrys and C in micromicrofarads. Hence,

$$f_r(mc) = \frac{159.1}{\sqrt{150 \times 169}} = \frac{159.1}{\sqrt{25,330}} = \frac{159.1}{159.1}$$

$$= 1 \text{ mc} = 1000 \text{ kc}$$

The resonant frequency of the circuit, thus, is 1000 kilocycles. At this frequency, the inductive reactance of the coil

$$X_L = 2 \pi f L = 6.283 \times 10^6 \times 150 \times 10^{-6} = 942 \text{ ohms}$$

The capacitive reactance of the capacitor

$$X_C = \frac{1}{2 \pi f C} = \frac{1}{6.283 \times 10^6 \times 169 \times 10^{-12}}$$

$$= 942 \text{ ohms.}$$

As expected, the inductive and capacitive reactances turn out to be *equal* at the resonant frequency. It is apparent from the formulas for capacitive and inductive reactance and from Fig. 158 that at **frequencies below resonance the capacitive reactance is greater than the inductive reactance** and, hence, the circuit is said to be **capacitive** at these frequencies. At **frequencies above resonance, in contrast, the inductive reactance exceeds the capacitive reactance** and, hence, the circuit is said to be **inductive above resonance.** This means that for frequencies below resonance the entire circuit can be replaced by an **equivalent capacitance** in series with a resistance, while for frequencies above resonance the circuit can be replaced by an **equivalent inductance** in series with a resistance.

Let us assume initially that the total series resistance, R, in the circuit, is 7.5 ohms. This includes the coil winding resistance, the resistance of the connecting leads, the leakage resistance of the capacitor, etc. With R = 7.5 ohms, the total current.

$$I = \frac{E}{R} = \frac{1}{7.5} = 0.1333 \text{ ampere}$$

The voltage drop across the resistance,

$$E_R = I R = 0.133 \times 7.5 = 1 \text{ volt}$$

The voltage drop across the inductance,

$$E_L = I X_L = 0.133 \times 942 = 125 \text{ volts (approx).}$$

And the voltage drop across the capacitance,

$$E_C = I X_C = 0.133 \times 942 = 125 \text{ volts}$$

Thus, we end up with the amazing situation that the voltage drops across both the inductance and capacitance are *125 times as great* as the voltage drop across the resistance or the *applied emf.* An emf of 1 volt produces coil and capacitor voltages of 125 volts each! This situation, which is typical of a resonant circuit, is pictured in vector diagram (a) of Fig. 160 below. Note that the voltage drops across the inductance and capacitance, being in *phase opposition,* cancel each other out completely, while the voltage drop across the resistance (I R) is *in phase* with the current and *equals the emf of the source* (each 1 volt).

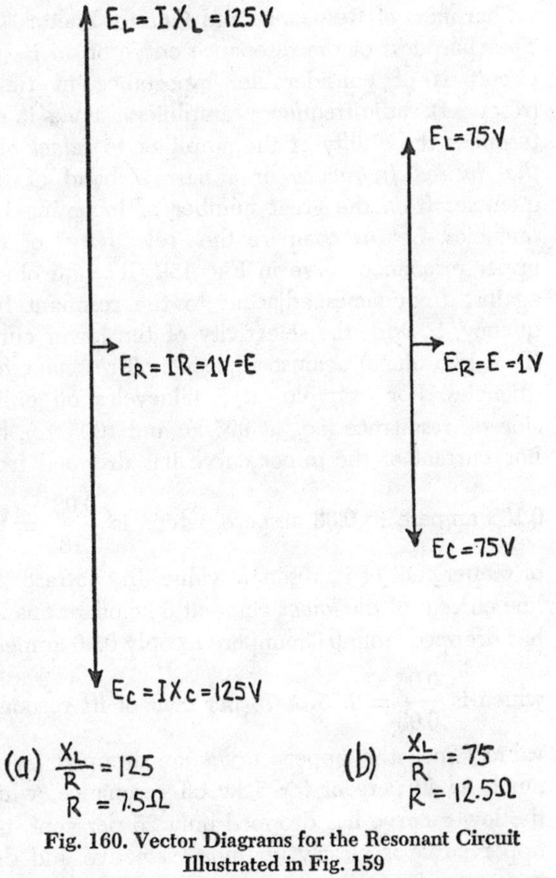

(a) $\dfrac{X_L}{R} = 125$ $R = 7.5\,\Omega$

(b) $\dfrac{X_L}{R} = 75$ $R = 12.5\,\Omega$

Fig. 160. Vector Diagrams for the Resonant Circuit Illustrated in Fig. 159

At the right of Fig. 159 we have plotted a graph of the total (line) *current* in the R-L-C circuit at left against the *frequency* of the applied voltage. Such a graph is known as a **resonance curve,** since it portrays the variation of the line current for frequencies near and at resonance. The upper curve has been drawn for a total circuit resistance (R) of 7.5 ohms, the case we have just discussed; while the lower curve applies for a circuit resistance of 12.5 ohms. Note how sharply the upper curve rises to its maximum current of 0.133 ampere at resonance, while the lower curve approaches its maximum point more gradually, with its sides (called "skirts") sloping out considerably. Because of the higher resistance, the lower curve has a maximum current at resonance of only $\frac{E}{R} = \frac{1}{12.5} = 0.08$ ampere, compared to 0.133 ampere for the upper curve. At frequencies considerably above or below resonance (about 25 kc on either side), the line current depends chiefly on the reactance and is practically independent of the circuit resistance. Both curves, therefore, merge for off-resonant frequencies of greater than about 25 kilocycles.

Sharpness of Resonance and Quality Factor "Q." The sharpness of the resonance curve of an R-L-C circuit is of considerable importance in tuned (resonant) radio-frequency amplifiers, since it determines the ability of the amplifier to *select only the desired frequency* or a narrow band of frequencies from the great number of incoming frequencies. Let us compare the "selectivity" of the upper resonance curve in Fig. 159 (R = 7.5 ohms) against frequencies adjacent to the resonant frequency, f_r, with the selectivity of the lower curve (R = 12.5 ohms) against adjacent, off-resonant frequencies. For example, at 5 kilocycles on either side off resonance (i.e., at 995 kc and 1005 kc), the line current of the *upper* curve has dropped from 0.133 ampere to 0.08 ampere, which is $\frac{0.08}{0.133} = 0.6$ or 60 per cent of its resonant value. In contrast, the line current of the *lower* curve at 5 kc off resonance has dropped from 0.08 ampere to only 0.06 ampere, which is $\frac{0.06}{0.08} = 0.75$ or 75 per cent of its resonant value. Since the upper curve has dropped by as much as 40 percent for 5 kc off resonance, while the lower curve has dropped only 25 per cent, the upper curve is clearly far more selective and discriminates far better against unwanted frequencies than the lower curve.

The sharpness of the resonance curve is determined by the ratio of the reactance of either the coil or the capacitor at the resonant frequency to the total resistance of the circuit. This ratio is known as the quality factor or "Q" of the circuit. Mathematically, the Q is defined:

$$Q = \frac{X_L}{R} = \frac{2\pi f L}{R}, \text{ or } Q = \frac{X_C}{R} = \frac{1}{2\pi f C R}$$

Since $X_L = X_C$ at resonance, both definitions result in the same value for Q. However, with most of the circuit resistance (R) being associated with the

winding resistance of the coil, you will usually find the term Q applied to the coil of a resonant circuit. Thus, for the lower resonance curve in Fig. 159, where R = 12.5 ohms,

$$Q = \frac{X_L}{R} = \frac{942}{12.5} = 75, \text{ or } Q = \frac{X_C}{R} = \frac{942}{12.5} = 75$$

The circuit Q also determines the ratio of the voltage drop across either the coil or the capacitor to the source voltage (E). This is easily shown:

The current $I = \frac{E}{R}$.

The voltage drop across the coil,

$$E_L = I X_L = \frac{E}{R} X_L$$

hence, $\frac{E_L}{E} = \frac{X_L}{R} = Q$; or $E_L = Q E$.

The voltage drop across the capacitor

$$E_C = I X_C = \frac{E}{R} X_C$$

hence, $\frac{E_C}{E} = \frac{X_C}{R} = Q$; or $E_C = Q E$.

Thus, in the lower resonance curve (Fig. 159), where R = 12.5 ohms, the line current $I = \frac{E}{R} = \frac{1}{12.5} = 0.08$ ampere, the voltage across the coil, $E_L = I X_L = 0.08 \times 942 = 75$ volts, and the voltage across the capacitor, $E_C = I X_C = 0.08 \times 942 = 75$ volts again. These results could have been obtained more simply by using the Q-factor:

The voltage across the coil, $E_L = Q E = 75 \times 1 = 75$ volts and the voltage across the capacitor, $E_C = Q E = 75 \times 1 = 75$ volts. The phase relations between E_L, E_C and E_R (=E) for the circuit of Fig. 159, when R = 12.5 ohms and Q = 75, are shown in the vector diagram of Fig. 160 (b). Note that the voltage across the resistance equals the applied emf (1 volt), as before.

EXAMPLE: A 10-microhenry coil with 20 ohms winding resistance is connected in an L-C circuit that resonates at 150 megacycles. What is the Q of the coil? If the applied emf is 5 millivolts, what is the coil voltage at resonance?

Solution: Q =

$$\frac{2\pi f L}{R} = \frac{6.283 \times 150 \times 10^6 \times 10 \times 10^{-6}}{20} = 471$$

the coil voltage $E_L = Q E = 471 \times 5 \times 10^{-3} = 2.36$ volts.

Since the coil voltage in a resonant circuit is "amplified" by the Q-factor over the applied voltage, it is evident why a high-Q coil is desirable.

You can perform a relatively simple experiment that will demonstrate clearly the basic facts of series resonance.

EXPERIMENT 25: The components needed for this experiment are: a 1000-ohm, 10-watt resistor; a 10-henry filter choke with about 200 ohms (d-c) winding resistance (or less); three paper capacitors, rated at 500 volts working voltage each, with values of 0.5, 0.2, and 0.3 microfarads (other combinations totaling up to these values are also acceptable); a 0-500 volt range A-C voltmeter; a 0-100 ma A-C milliammeter; and a 200-ma fuse to protect the components. Hook the components in series with each other and the 120-volt, 60-cycle a-c power line, as illustrated in Fig. 161. Initially, place only the 0.5-microfarad capacitor in series with the 10-henry choke and the 1000-ohm resistor.

We shall approach resonance by changing the capacitance, since the frequency of the power line (60 cycles) is not under our control. As you can easily check, the inductive reactance of the coil is about 3770 ohms. With the 0.5-μf capacitor in the circuit, the capacitive reactance turns out to be about 5300 ohms, so that the net reactance is 3770 $-$5300 $= -1530$ ohms *capacitive*. Since the total resistance is 1200 ohms (i.e., the 1000-ohm resistor plus 200 ohms coil winding resistance), the total impedance turns out to be $Z = \sqrt{(1200)^2 + (-1530)^2} = 1940$ ohms. Hence, the line current measured on the

0-100 milliammeter should be about $\frac{120}{1940} = 0.062$ ampere or 62 ma. Placing the 0-500 volt A-C voltmeter across the choke, you should measure a voltage of about $0.062 \times 3770 = 234$ volts coil voltage. Next place the voltmeter across the capacitor. It should read a voltage drop of about $0.062 \times 5300 = 328$ volts capacitor voltage. Finally, place the voltmeter across the 1000-ohm resistor and read a voltage of about $0.062 \times 1000 = 62$ volts. Although the choke coil and capacitor voltages each exceed the line voltage considerably, the circuit is *not at resonance*, since the net reactance ($X_L - X_C$) is *not* equal to zero.

Now place the 0.2-microfarad capacitor in parallel with the 0.5-μf capacitor to obtain a total capacitance of about 0.7 microfarads. The capacitive reactance is now $\frac{1 \times 10^6}{6.28 \times 60 \times 0.7} = 3770$ ohms (approximately) or the *same as the inductive reactance* (which is unchanged). By the definition of series resonance ($X_L = X_C$), the circuit is now at resonance. The line current indicated on the milliammeter will rise to the value of E/R or 120/1200 = 0.1 ampere or 100 ma. (The total resistance R = 1000 + 200 = 1200 ohms.) The "Q" of the resonant circuit is $\frac{X_L}{R} = \frac{3770}{1200} = 3.14$, which is not very high. (Much higher Q's could be produced easily by lowering the value of the resistor, but this would

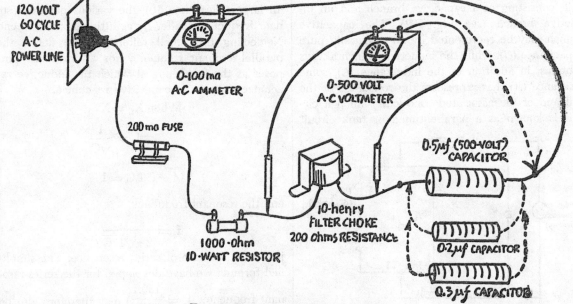

120 VOLT 60 CYCLE A·C POWER LINE

0-100 ma A·C AMMETER

200 ma FUSE

1000-ohm 10-WATT RESISTOR

0-500 VOLT A·C VOLTMETER

10-henry FILTER CHOKE 200 ohms RESISTANCE

0.5 μf (500-VOLT) CAPACITOR

0.2 μf CAPACITOR

0.3 μf CAPACITOR

Fig. 161. Experiment 25: Series Resonance

result in a large line current and excessive voltages, which would damage the components or blow the fuse.) The voltage drops across the choke coil and the capacitor rise to Q times the line voltage at resonance, or to $3.14 \times 120 = 377$ volts each. The voltage drop across the resistor should read about $0.1 \times 1000 = 100$ volts. Do not keep the circuit in resonance too long to avoid overheating and possible damage to the components.

Finally, let us take the circuit out of resonance by connecting an *excess* capacitance. When the 0.3-microfarad capacitor is added in parallel with the 0.5 and 0.2-μf capacitors, the total capacitance will be 1 microfarad, and the capacitive reactance turns out to be about 2,650 ohms. The net reactance is now $3770 - 2650 = 1120$ ohms *inductive* and, hence, the entire circuit behaves like an inductance. The total impedance in this case is

$$\sqrt{(1200)^2 + (1120)^2} = 1640$$

ohms and the line current indicated on the milliammeter should be about $\dfrac{120}{1640} = 0.073$ amp. or 73 ma.

Note that the current has dropped considerably from its maximum value at resonance. The voltage across the choke coil should now read $0.073 \times 3770 = 276$ volts and that across the capacitor, $0.073 \times 2650 = 194$ volts, each being considerably less than the values measured at resonance. The voltage across the resistor should read $0.073 \times 1000 = 73$ volts.

PARALLEL RESONANCE

Resonance also can be attained in a parallel circuit, consisting of a capacitive branch and an inductive branch (see Fig. 162). The capacitive branch may be represented by an essentially pure capacitance (C), while the inductive branch always contains, in addition to the inductance (L), some resistance (R) that represents the resistance of the coil and of the associated conductors. Such a circuit is known as a parallel-tuned or tank circuit

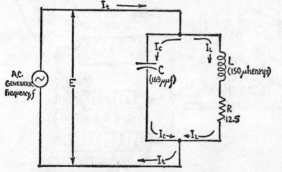

Fig. 162. Parallel-Resonant (Tank) Circuit

and it is one of the most widely used devices in radio transmitters and receivers.

Parallel-Resonant Frequency. When an a-c voltage (E) is impressed across a parallel-tuned (tank) circuit, electrical resonance occurs at approximately the same frequency at which the circuit would "naturally" oscillate when excited. Analogous to the mechanical oscillating system we have discussed, a parallel-resonant circuit just draws sufficient energy from the a-c supply to overcome its internal (resistance) losses. This, indeed, is the fundamental meaning of resonance: at the resonant frequency the external energy supply releases just enough energy with the proper timing to sustain the natural self-oscillations of the tank circuit. There are, however, at least **three definitions** as to just what this parallel-resonant frequency of a tank circuit is. These definitions of parallel resonance are:

1. The frequency at which the line current is *in phase* with the impressed line voltage. This is the condition of **unity power factor**.

2. The frequency at which the **inductive reactance equals the capacitive reactance** $(X_L = X_C)$, as in the case of series resonance.

3. The frequency at which the **impedance of the tank circuit is a maximum**, or equivalently, the **line current is a minimum**.

When the resistance of the tank circuit is *high* and the *ratio* of (inductive or capacitive) *reactance to the resistance* (i.e. the "Q") is *low*, these definitions lead to parallel-resonant frequencies that differ slightly with each other and the frequency of natural oscillations. Fortunately, when the Q of the circuit is at all large (more than 10), the various parallel-resonant frequencies differ by as little as one per cent. Neglecting this small difference, we define the **parallel-resonant frequency** for all practical purposes as the **frequency at which the inductive reactance equals the capacitive reactance.**

When $X_L = X_C$

$$wL = \frac{1}{wC}$$

$$4\pi^2 f^2 LC = 1$$

and the resonant frequency,

$$f_r = \frac{1}{2\pi\sqrt{LC}}$$

just as in the case of series resonance. The simplified formula we have developed for the series-resonant frequency $\left(f_r = \dfrac{159.1}{\sqrt{LC}}\right)$ may therefore also be applied to parallel resonance.

To get an idea of what happens at parallel resonance, let us modify our circuit from the experiment in series resonance slightly to make it a parallel-resonant circuit.

EXPERIMENT 26: We shall use the same components as for Experiment 25, except that we omit the 1000-ohm resistor, since we have no longer any need to limit the current and the voltage drops and want to attain a circuit Q that is as high as possible. Initially, connect the 10-henry, 200-ohm choke in parallel with the 0.5-microfarad capacitor and connect a lead from each side of the parallel combination to the 120-volt, 60-cycle a-c power line, as illustrated in Fig. 163. You will need the 0-500 volt a-c voltmeter only once, to measure the line voltage, which is also the voltage across the parallel L-C combination. After you have recorded the line-voltage value, you can lay the a-c voltmeter aside. Since we have to measure the currents in various parts of the circuit, make provisions to insert the 0-100 ma, a-c milliammeter into the line, as well as into the inductive and capacitive branches. You can break the hook-up wires at these current-measuring points to permit connecting the meter into the circuit, but be sure the wires are twisted together and make contact when the ammeter is not inserted.

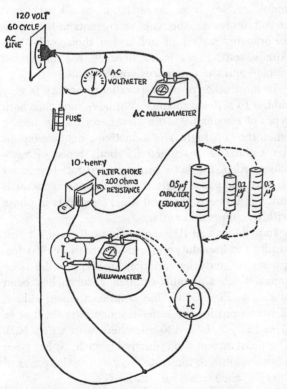

Fig. 163. Experiment 26: Parallel Resonance

Again we shall approach (parallel) resonance by varying the capacitance. With $C = 0.5$ μf, the (negative) capacitive reactance

$$X_C = -\frac{1}{6.28 \times 60 \times 0.5 \times 10^{-6}} = -5300 \text{ ohms,}$$

as before, while the inductive reactance $X_L = 6.283 \times 60 \times 10 = 3770$ ohms. With only the 200 ohms winding resistance present, the Q of the circuit is raised to $\frac{X_L}{R} = \frac{3770}{200} = 18.85$ or about 19.

The *impedance* of the inductive branch,

$$Z_L = \sqrt{R^2 + X_L^2} = \sqrt{(200)^2 + (3770)^2} = 3780 \text{ ohms (approximately).}$$

Hence, the current through the inductive branch,

$$I_L = \frac{E}{Z_L} = \frac{120}{3780} = 0.0318 \text{ ampere. The milliam-}$$

meter, thus, should indicate an inductive current of about 32 milliamps. The current through the capacitive branch, $I_C = \frac{E}{X_C} = \frac{120}{5300} = 0.0227$ amp. or about 23 ma on the meter. Finally, the impedance of a parallel circuit with branch impedances Z_L and Z_C:

$$Z = \frac{Z_L \times Z_C}{Z_L + Z_C}, \text{ and since } Z_C = X_C, Z = \frac{Z_L X_C}{Z_L + X_C}.$$

Note that the denominator of this expression is the total *series* impedance of the tank circuit, and since Z_L is primarily inductive reactance in this case (R being very small), we may substitute the *net reactance* in the denominator with very little error resulting. Hence, approximately,

$$Z = \frac{Z_L X_C}{X} = \frac{Z_L X_C}{-(X_C - X_L)} = \frac{3780 \times 5300}{5300 - 3770} =$$

$$\frac{3780 \times 5300}{1530} = 13,100 \text{ ohms}$$

(where we have disregarded the — sign, since we only want the magnitude). The total line current is

therefore $I_t = \frac{E}{Z} = \frac{120}{13100} = 0.00916$ ampere. Thus, the milliammeter, when inserted into the main line (Fig. 163), should read about 9 milliamperes line current.

Now connect the 0.2-microfarad capacitor in parallel with the 0.5-microfarad capacitor to obtain a total capacitance of about 0.7 microfarad. This value, you will recall, produced series resonance in Experiment 25; and since the parallel-resonant frequency is the same, it should also produce parallel resonance in the present setup. The inductive and capacitive reactances now equal 3770 ohms each, as in the case of series resonance (experiment 25). The impedance of the inductive branch is 3780 ohms, as before, and the milliammeter should again read about 32 ma in the inductive branch. The current in the capacitive branch, $I_c = \dfrac{120}{3770} = 0.0318$ ampere, or about the same as the current in the inductive branch (32 ma). The total impedance of the tank circuit is again

$$Z = \frac{Z_L X_C}{Z_L + X_C} = \frac{Z_L X_C}{Z_{series}},$$

or stated in words, the product of the branch impedances divided by the total *series* impedance. Since the inductive and capacitive reactances cancel in this case (each being 3770 ohms), the total series impedance is simply the circuit *resistance* of 200 ohms. Hence, the total impedance

$$Z = \frac{Z_L \times X_C}{Z_{series}} = \frac{3780 \times 3770}{200} = 71{,}000 \text{ ohms}$$

Thus, the total impedance at parallel resonance has *risen* to over five times its non-resonant value of 13,200 ohms. The line current, consequently, has decreased to a much *lower value* of

$$I_t = \frac{E}{Z} = \frac{120}{71{,}000} = 0.00169 \text{ ampere,}$$

or less than one-fifth of its former value. When inserted into the main line, the milliammeter will now read a barely perceptible current between 1 and 2 ma.

Finally, complete the experiment by connecting the 0.3-μf capacitor in parallel with 0.7-μf combination, obtaining a total capacitance of about 1 microfarad. The inductive impedance is 3780 ohms, as before, and the current in the inductive branch should read again about 32 ma. The capacitive reactance, however, is now

$$X_C = \frac{1}{6.28 \times 60 \times 1 \times 10^{-6}} = 2650 \text{ ohms}$$

and, hence, the current in the capacitive branch should read $\dfrac{120}{2650} = 0.0453$ ampere, or about 45 ma. The net reactance is now $X_L - X_C = 3770 - 2650 = 1120$ ohms, and hence the *series* impedance of the tank circuit is $Z_{series} = \sqrt{(200)^2 + (1120)^2} = 1138$ ohms. Thus, the total parallel impedance

$$Z = \frac{Z_L \times X_C}{Z_{ser}} = \frac{3780 \times 2650}{1138} = 8800 \text{ ohms,}$$

or a value of about *one-eighths* (approx.) of the resonant impedance. The line current, accordingly,

has increased to about $\dfrac{120}{8{,}800} = 0.0136$ ampere, or 13.6 ma. We conclude that the impedance of a parallel-resonant circuit is a maximum, while the line current is a minimum.

Parallel-Resonant Circuits. Experiment 26 has demonstrated clearly that the behavior of a parallel-resonant circuit is radically different from that of a series-resonant circuit in at least two aspects: 1. the impedance of a parallel-resonant circuit is a maximum, while that of a series-resonant circuit is a minimum; 2. the total (line) current of a parallel-resonant circuit is a minimum, while that of a series-resonant is a maximum. There is one other important difference between these two types of resonance. We have seen that in a series-resonant circuit the individual voltage drops across the inductance and capacitance may be large, but the sum of the reactive voltage drops must equal zero, and, moreover, the voltage drop across the resistance equals the impressed (supply) voltage. Since in a parallel circuit the same, constant supply voltage is impressed across all branches, there can be no voltage rise across any branch at resonance. The individual currents in the inductive and capacitive branches, however, may be quite large at resonance, though their vector sum—which is the line current—is very small, as we have seen. In a parallel-resonant circuit, moreover, the reactive currents in the parallel branches are equal and, hence, their vector sum equals zero. (This is so because the impressed voltage and the reactances are the same.)

In most other respects, parallel resonance is very similar to series resonance. We have seen that both types of circuits have the same resonant frequency, when the Q is high. The inductive reactance equals the capacitive reactance for both types of resonance. Also, for both series and parallel-resonant circuits, the phase angle is zero, the power factor is unity and, hence, the total (line) current is in phase with the impressed voltage.

Impedance and Resonance Curve. It is of interest to plot the impedance of a tank circuit against frequency to study its variation around parallel resonance. Such a parallel-resonance curve has been plotted in Fig. 164 for the same component values, as were used for the series-resonant circuit; that is, C = 169 μμf, L = 150 microhenrys (see Fig. 162). The resistance of the inductive branch, R, has been made variable to illustrate the relative sharpness of the resonance curve for various Q's.

Note that the variation of impedance with frequency of a tank circuit near *parallel* resonance is similar to the variation of the line current with frequency of an R-L-C series circuit near *series* resonance (See Fig. 159.) The impedance of a tank circuit is a **maximum at the parallel-resonant frequency** (1000 kc, in this case). Again the sharpness of the resonance curve depends on the Q of the circuit; that is, on the ratio of the (inductive or capacitive) reactance at resonance to the total circuit resistance. For Q's that are low, the impedance curve rises moderately at resonance; for a high circuit Q, the impedance rises sharply at resonance; for an "infinite" Q (zero resistance), the resonant impedance—theoretically—becomes infinite. Since there is always some resistance, this is not realizable in practice.

Fig. 164 also shows a graph of the phase angle of the impedance against frequency, near resonance. The phase angle is seen to be **lagging below resonance, zero at resonance, and leading above resonance.** Accordingly, a parallel L-C (tank) circuit

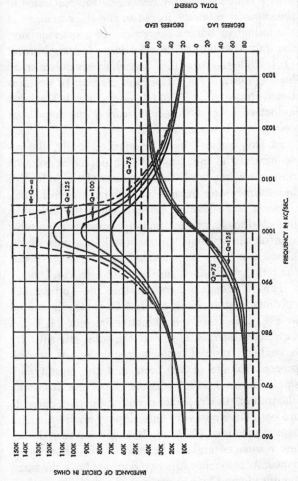

Fig. 164. Variation of Magnitude and Phase Angle of Tank Circuit Impedance with Frequency and Circuit Q.

is **inductive below resonance, resistive at resonance,** and **capacitive above resonance.**

Magnitude of Impedance. We can easily derive an approximate expression for the impedance of a parallel-resonant circuit. The impedance of the two parallel branches of a tank circuit (Fig. 162)

$$Z = \frac{Z_L Z_C}{Z_L + Z_C} = \frac{Z_L Z_C}{Z_{series}},$$

where Z_L is the impedance of the inductive branch, Z_C is the impedance of the capacitive branch, and Z_{series} ($= Z_L + Z_C$) is the total series impedance of the two branches added together. Since there is no resistance in the capacitive branch, $Z_C = X_C$. Moreover, when the circuit Q is at all large (greater than 10), the impedance of the inductive branch approximately equals the inductive reactance ($Z_L = X_L$ approx.), so that we can write

$$Z = \frac{X_L X_C}{Z_{series}}$$

At the **parallel-resonant frequency:**

$$X_L = X_C$$

and the series impedance, $Z_{series} = R$ (i.e., the circuit resistance). Hence, at resonance,

$$Z = \frac{X_L X_L}{R} = Q X_L = Q w L \text{ (since } Q = \frac{X_L}{R})$$

Thus, the **impedance of a parallel-resonant circuit** is approximately Q *times the inductive* (or capacitive) *reactance* at resonance.

We also have the expression, $Q = \frac{X_C}{R} = \frac{1}{w C R}$.

Substituting this relation for Q in the impedance formula, we finally obtain:

$$Z = Q w L = \frac{w L}{w C R} = \frac{L}{C R}$$

Hence, the **resonant impedance** of a tank circuit is also approximately **equal to the inductance divided by the product of capacitance and resistance.** The impedance is **purely resistive,** since there is *no phase angle.*

EXAMPLE: In the circuit of Fig. 162, L = 150 microhenrys; C = 169 $\mu\mu$f; and let R = 12.5 ohms. What is the Q and the impedance of the circuit at resonance?

Solution: We have already previously determined the resonant frequency, $f_r = \frac{159.1}{\sqrt{L C}} = \frac{159.1}{\sqrt{150 \times 169}}$ = 1 mc = 1000 kilocycles, and the inductive reactance, $X_L = 2\pi f L = 6.283 \times 10^6 \times 150 \times 10^{-6} =$ 942 ohms. Hence, the $Q = \frac{X_L}{R} = \frac{942}{12.5} = 75$ (ap-

proximately). Thus, the resonant impedance
Z = Q X$_L$ = 75 × 942 = 70,500 ohms

(approximately).

Equivalently,

$$Z = \frac{L}{C\,R} = \frac{150 \times 10^{-6}}{169 \times 10^{-12} \times 12.5} = 71,000 \text{ ohms.}$$

Both these values check well with the impedance value of the resonance curve for Q = 75 and a frequency of 1000 kc.

Current and Voltage Relations. Since the impedance of a tank circuit approaches a maximum at resonance, the line current approaches a minimum. When a curve is plotted of the variation of the line current with frequency in a parallel-tuned circuit, its shape turns out to be the inverse of the impedance curve of Fig. 164. A typical plot of line current vs. frequency is shown in Fig. 165.

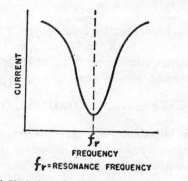

Fig. 165. Variation of Line Current with Frequency
near Parallel Resonance

The phase relationships between the currents and the impressed voltage at, below and above resonance are also of interest. At resonance, the total (line) current is resistive and is given by

$$I_t = \frac{E}{Z} = \frac{E}{Q\,X_L} = \frac{E}{Q\,w\,L}$$

The capacitive and inductive currents are much larger than the total current and they are nearly but not quite (because of the resistance) 180 degrees out of phase with each other. Since X$_L$ = X$_C$ at resonance, the currents are approximately *equal in magnitude* and are given by the relation

$$I_L = I_C \text{ (approx.)} = \frac{E}{X_L} = \frac{E}{w\,L}$$

By comparing with the previous relation for I$_t$, we also see that

I$_L$ = I$_C$ = Q I$_t$ (approximately)

The **phase relations at resonance** are shown in Fig. 166a. You can see that the capacitive current is large and leads the voltage by 90°; the inductive

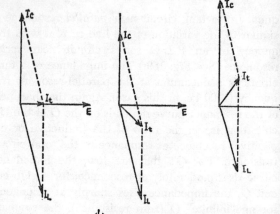

'(a) AT RESONANCE   (b) BELOW RESONANCE   (c) ABOVE ?

Fig. 166. Vector Diagrams of Currents and Voltage (a) at resonance, (b) below resonance and (c) above resonance

current is equally large and lags behind the voltage by somewhat less than 90° because of the resistance in the inductive branch. The vector sum of these currents is the total (line) current, I$_t$, which is seen to be very small, but *in phase with the voltage.*

At frequencies below resonance the reactance of the **capacitive** branch *increases* and the reactance of the inductive branch *decreases*. As a consequence, the inductive branch draws a *large lagging current* (I$_L$), while the capacitive branch draws only a *small leading current* (I$_C$). The total current, I$_t$, which is the vector sum of I$_L$ and I$_C$, thus, is a fairly large **inductive current** that *lags behind the impressed voltage,* as shown in Fig. 166b.

At frequencies above resonance, finally, the reactance of the **capacitive** branch *decreases* from its resonant value, while the reactance of the **inductive branch increases.** As a result, the capacitive branch draws a **large leading current,** while the inductive branch draws a smaller **lagging current.** The vector sum of the currents, or the total line current, I$_t$, hence, is a fairly large *capacitive current that leads the impressed voltage,* as illustrated in Fig. 166c.

The relationship between the line and branch currents for the circuit of Fig. 162 is also illustrated in Fig. 167. Here the inductive current (I$_L$), the capacitive current (I$_C$) and the total current (I$_t$) have been plotted against frequency for an impressed voltage of 100 volts and the circuit constants shown in Fig. 162. As is evident from the illustration, the capacitive and inductive currents are both large and vary only slightly for frequencies near resonance (1000 kc). I$_L$ is greater than I$_C$ *below* resonance and the reverse is true *above* resonance. At resonance, the two currents have the same magnitude and being nearly 180° out of phase, their vector sum adds up to a very small, resultant line current, I$_t$. With the current being a minimum at

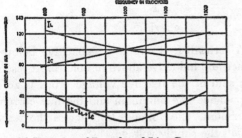

Fig. 167. Variation of Branch and Line Currents near Parallel Resonance

resonance, the impedance reaches its maximum, of course.

Summary of Conditions at Parallel Resonance. Let us summarize the conditions prevailing at the parallel-resonant frequency of a tank circuit, when the Q is above 10.

Reactances: $X_L = X_C$

$$wL = 1/wC$$

(inductive reactance equals capacitive reactance)

Resonant Frequency:

$$f_r = \frac{1}{2\pi\sqrt{LC}} = \frac{159.1}{\sqrt{LC}} \times 10^{-3}$$

Branch Currents: $I_L = I_C = Q I_t$

(The branch currents are *equal* to Q times the line current)

Total (Line) Current: $I_t = \dfrac{E}{Z} = \dfrac{E}{QwL}$

(line current is a *minimum*)

Total Impedance: $Z = Q X_L = Q w L = \dfrac{L}{CR}$

(impedance is a *maximum*)

Phase Angle: $\theta = 0$

(line current is *in phase* with applied voltage)

Power Factor: P.F. $= 1$

(power factor is *unity*)

EXAMPLE: A coil with an inductance of 160 microhenrys and an effective resistance of 20 ohms is connected in parallel with a 250-$\mu\mu$f capacitor. The tank circuit is then connected to a 20-volt, variable-frequency, a-c source. What are the resonant frequency, the Q, the total circuit impedance at resonance, the line current and the branch currents at resonance?

Solution: The resonant frequency, $f_r =$

$$\frac{159.1}{\sqrt{160 \times 250}}$$

$= 0.796$ mc $= 796$ kc; the circuit $Q = \dfrac{wL}{R} =$

$$\frac{6.283 \times 796 \times 10^3 \times 160 \times 10^{-6}}{20} = 40;$$ the total

impedance $Z = Q w L = 40 \times 800 = 32,000$ ohms

(approximately); or $Z = \dfrac{L}{CR} = \dfrac{160 \times 10^{-6}}{250 \times 10^{-12} \times 20}$

$= 32,000$ ohms; the line current,

$$I_t = \frac{E}{Z} = \frac{20}{32,000} = 0.625 \times 10^{-3} = 625 \text{ micro-}$$

amps; the branch currents, $I_L = I_C = Q I_t = 40 \times 0.625 \times 10^{-3} = 25 \times 10^{-3} = 25$ milliamps. (approximately)

Practice Exercise No. 14

1. What is meant by "resonance" in a physical sense? Cite examples.

2. Explain the analogy between a mechanical and electrical oscillating system, such as a spring pendulum and a tank circuit.

3. Fill in the analogous terms of a mechanical or electrical oscillating system below:
   (a) Inductance _____
   (b) $\dfrac{1}{\text{Elasticity}}$ _____
   (c) Displacement _____
   (d) Energy of Electric Field _____
   (e) Kinetic Energy _____

4. Define the following terms: a. Oscillator; b. Forced oscillations; c. Natural frequency; d. Resonant frequency.

5. For a given impressed force (or voltage), at what frequency is the amplitude of oscillations in an oscillating system greatest? When is minimum energy taken from the supply?

6. Define series resonance in an R-L-C circuit and derive the resonant frequency.

7. Prove that for series resonance, the power factor is unity, the phase angle is zero, and the total current, I = E/R.

8. What should the inductance of a coil be so that it produces resonance at 60 cps when connected in series with a 20-μf capacitor?

9. A 160-microhenry inductor with a winding resistance of 20 ohms is connected in series with a 250-$\mu\mu$f capacitor and an a-c voltage of 0.5 volt is applied to the circuit. What are (a) the resonant frequency of the circuit; (b) the resonant line current; (c) the magnitude of the voltage drop across the coil and (d) the voltage drop across the capacitor?

10. What is the Q of the circuit in the previous problem?

11. State the three definitions of parallel resonance and give the resonant frequency, when Q is greater than 10.

12. Describe the characteristics of a parallel-resonant circuit and draw vector diagrams of the line and branch currents at resonance, below resonance, and above resonance.

13. Derive the impedance of a parallel-resonant circuit in terms of (a) the Q and the inductive reactance; (b) L, C, and R. What is the phase angle of the impedance at resonance?

14. A 0.1-henry coil with 10 ohms resistance is connected in parallel with a 70-microfarad capacitor of negligible resistance. At what (approximate) frequency does the circuit act like a non-inductive resistance and what is the approximate value of this resistance?

15. A 300-microhenry coil with 18 ohms resistance is connected in parallel with a loss-less capacitor and the combination is connected to a 10-volt a-c source. If resonance is to occur at 450 kc, determine (a) the capacitance needed for resonance; (b) the circuit Q; (c) the resonant impedance; (d) the line current at resonance, and (e) the current through the inductance at resonance.

SUMMARY

An electrical oscillating system is analogous to a mechanical oscillator (pendulum) and must have the following elements:

1. **inductance** to resist any change in the current—analogous to the **inertia** of a mass (weight of pendulum); 2. **capacitance** to store electric energy (charge)—analogous to the **inverse of elasticity** (displacement of a spring); and 3. a source of **energy** (**voltage**) to excite the oscillations with the proper timing (frequency)—analogous to **mechanical force**. An oscillating system usually also has **resistance**—analogous to **mechanical friction**.

Impressing an alternating current of a certain frequency upon an R-L-C oscillating system is analogous to **forcing a mechanical oscillator** (pendulum) to vibrate at a *predetermined* frequency.

The **amplitude of forced oscillations** in a mechanical or electrical oscillating system is greatest, when the **frequency of the impressed force or voltage equals the natural frequency of oscillations**. Either system is then said to be in resonance and draws just **sufficient energy** from the supply to overcome its internal (resistance) losses.

Series resonance takes place in a series-connected R-L-C circuit, when the **inductive reactance equals the capacitive reactance** ($X_L = X_C$), so that the net reactance is zero. The circuit is then **purely resistive**, has **unity power factor**, and the line current is in phase with the impressed voltage. ($Z = R$;

$I = \dfrac{E}{R}$; $\theta = 0$; P.F. = 1). The series-resonant fre-

quency, $f_r = \dfrac{1}{2\pi\sqrt{L\,C}}$.

A series-R-L-C circuit is **capacitive below resonance** and **inductive above resonance**. At resonance, the sum of the reactive voltage drops is zero, the impedance is a minimum and the line current is a maximum. The voltage drop across the resistance equals the supply voltage.

The **quality factor "Q"** is the **ratio of the reactance of the coil or that of the capacitor at the resonant frequency to the total resistance**.

$(Q = \dfrac{X_L}{R} = \dfrac{w\,L}{R} = \dfrac{X_C}{R} = \dfrac{1}{w\,C\,R})$. The circuit Q

determines the **sharpness of the resonance curve**.

The voltage drop across the coil of a series-resonant circuit equals the voltage drop across the capacitor and both are approximately equal to Q times the supply voltage ($E_L = E_C = Q\,E$).

Three definitions of parallel resonance are:

1. frequency at which the **line current is in phase with impressed voltage**. (Condition of unity power factor).

2. frequency at which the **inductive reactance equals the capacitive reactance** ($X_L = X_C$).

3. frequency at which the **impedance of the circuit is a maximum and the line current is a minimum**.

When the resistance of a parallel-connected R-L-C (tank) circuit is low and the Q is fairly high (greater than 10), the three definitions lead to the

same parallel-resonant frequency, $f_r = \dfrac{1}{2\pi\sqrt{L\,C}}$.

The impedance of a parallel-resonant circuit is a maximum, the line current is a minimum and in phase with the impressed voltage (phase angle is zero, power factor is unity). At resonance, the sum of the reactive currents in the parallel branches equals zero.

A tank circuit has a **lagging** phase angle below resonance and, hence, is **inductive below resonance**; above resonance the phase angle is **leading** and the circuit is **capacitive**.

At parallel resonance:

impedance $Z = Q\,X_L = Q\,w\,L = \dfrac{L}{C\,R}$;

total (line) current, $I_t = \dfrac{E}{Z} = \dfrac{E}{Q\,w\,L}$

branch currents, $I_L = I_C = Q\,I_t$.

APPENDIX I

ELECTRICAL DATA

TABLE VII—COPPER-WIRE TABLE—STANDARD ANNEALED COPPER: *American wire gauge (B & S)*

gauge no	diam-eter, mils	cross section circular mils	cross section square inches	ohms per 1,000 ft at 20° C (68° F)	lb per 1,000 ft	ft per ohm at 20° C ft per lb	ft per ohm at 20° C (68° F)	ohms per lb at 20° C (68° F)
0000	460.0	211,600	0.1662	0.04901	640.5	1.561	20,400	0.00007652
000	409.6	167,800	0.1318	0.06180	507.9	1.968	16,180	0.0001217
00	364.8	133,100	0.1045	0.07793	402.8	2.482	12,830	0.0001935
0	324.9	105,500	0.08289	0.09827	319.5	3.130	10,180	0.0003076
1	289.3	83,690	0.06573	0.1239	253.3	3.947	8,070	0.0004891
2	257.6	66,370	0.05213	0.1563	200.9	4.977	6,400	0.0007778
3	229.4	52,640	0.04134	0.1970	159.3	6.276	5,075	0.001237
4	204.3	41,740	0.03278	0.2485	126.4	7.914	4,025	0.001966
5	181.9	33,100	0.02600	0.3133	100.2	9.980	3,192	0.003127
6	162.0	26,250	0.02062	0.3951	79.46	12.58	2,531	0.004972
7	144.3	20,820	0.01635	0.4982	63.02	15.87	2,007	0.007905
8	128.5	16,510	0.01297	0.6282	49.98	20.01	1,592	0.01257
9	114.4	13,090	0.01028	0.7921	39.63	25.23	1,262	0.01999
10	101.9	10,380	0.008155	0.9989	31.43	31.82	1,001	0.03178
11	90.74	8,234	0.006467	1.260	24.92	40.12	794	0.05053
12	80.81	6,530	0.005129	1.588	19.77	50.59	629.6	0.08035
13	71.96	5,178	0.004067	2.003	15.68	63.80	499.3	0.1278
14	64.08	4,107	0.003225	2.525	12.43	80.44	396.0	0.2032
15	57.07	3,257	0.002558	3.184	9.858	101.4	314.0	0.3230
16	50.82	2,583	0.002028	4.016	7.818	127.9	249.0	0.5136
17	45.26	2,048	0.001609	5.064	6.200	161.3	197.5	0.8167
18	40.30	1,624	0.001276	6.385	4.917	203.4	156.6	1.299
19	35.89	1,288	0.001012	8.051	3.899	256.5	124.2	2.065
20	31.96	1,022	0.0008023	10.15	3.092	323.4	98.50	3.283
21	28.46	810.1	0.0006363	12.80	2.452	407.8	78.11	5.221
22	25.35	642.4	0.0005046	16.14	1.945	514.2	61.95	8.301
23	22.57	509.5	0.0004002	20.36	1.542	648.4	49.13	13.20
24	20.10	404.0	0.0003173	25.67	1.223	817.7	38.96	20.99
25	17.90	320.4	0.0002517	32.37	0.9699	1,031.0	30.90	33.37
26	15.94	254.1	0.0001996	40.81	0.7692	1,300	24.50	53.06
27	14.20	201.5	0.0001583	51.47	0.6100	1,639	19.43	84.37
28	12.64	159.8	0.0001255	64.90	0.4837	2,067	15.41	134.2
29	11.26	126.7	0.00009953	81.83	0.3836	2,607	12.22	213.3
30	10.03	100.5	0.00007894	103.2	0.3042	3,287	9.691	339.2
31	8.928	79.70	0.00006260	130.1	0.2413	4,145	7.685	539.3
32	7.950	63.21	0.00004964	164.1	0.1913	5,227	6.095	857.6
33	7.080	50.13	0.00003937	206.9	0.1517	6,591	4.833	1,364
34	6.305	39.75	0.00003122	260.9	0.1203	8,310	3.833	2,168
35	5.615	31.52	0.00002476	329.0	0.09542	10,480	3.040	3,448
36	5.000	25.00	0.00001964	414.8	0.07568	13,210	2.411	5,482
37	4.453	19.83	0.00001557	523.1	0.06001	16,660	1.912	8,717
38	3.965	15.72	0.00001235	659.6	0.04759	21,010	1.516	13,860
39	3.531	12.47	0.000009793	831.8	0.03774	26,500	1.202	22,040
40	3.145	9.888	0.000007766	1,049.0	0.02993	33,410	0.9534	35,040

TABLE VIII—COPPER-WIRE TABLE—ENGLISH AND METRIC UNITS

Amer wire gauge AWG (B&S)	Birm wire gauge BWG	imperial or British std SWG (NBS)	diam in inches	English units weight lbs per wire mile	resistance ohms per wire mile 20° C (68° F)	diam in mm	metric units weight kg per wire km	resistance ohms per wire km 20° C (68° F)
—	—	—	.1968	618	1.415	5.0	174.0	.879
—	—		.1940	600	1.458	4.928	169.1	.905
		6	.1920	589.2	1.485	4.875	166.2	.922
5	—	—	.1855	550	1.590	4.713	155.2	.987
	7		.1819	528.9	1.654	4.620	149.1	1.028
			.1800	517.8	1.690	4.575	146.1	1.049
		7	.1771	500	1.749	4.5	141.2	1.086
			.1762	495.1	1.769	4.447	140.0	1.098
—	—	—	.1679	450	1.945	4.260	127.1	1.208
6	8		.1650	435.1	2.011	4.190	123.0	1.249
			.1620	419.5	2.086	4.115	118.3	1.296
		8	.1600	409.2	2.139	4.062	115.3	1.328
—	—	—	.1582	400	2.187	4.018	113.0	1.358
—	—	—	.1575	395.3	2.213	4.0	111.7	1.373
7	9		.1480	350.1	2.500	3.760	98.85	1.552
			.1443	332.7	2.630	3.665	93.78	1.634
		9	.1440	331.4	2.641	3.658	93.40	1.641
—	—	—	.1378	302.5	2.892	3.5	85.30	1.795
—	—	—	.1370	300	2.916	3.480	84.55	1.812
	10		.1341	287.0	3.050	3.405	80.95	1.893
8			.1285	263.8	3.317	3.264	74.37	2.061
		10	.1280	261.9	3.342	3.252	73.75	2.077
—	—	—	.1251	250	3.500	3.180	70.50	2.173
9			.1181	222.8	3.930	3.0	62.85	2.440
			.1144	209.2	4.182	2.906	58.98	2.599
—	—	—	.1120	200	4.374	2.845	56.45	2.718
	12		.1090	189.9	4.609	2.768	53.50	2.862
		12	.1040	172.9	5.063	2.640	48.70	3.144
*10			.1019	165.9	5.274	2.588	46.77	3.277
—	—	—	.0984	154.5	5.670	2.5	43.55	3.520
—	—	—	.0970	150	5.832	2.460	42.30	3.620
	*14		.0830	110.1	7.949	2.108	31.03	4.930
*12			.0808	104.4	8.386	2.053	29.42	5.211
		14	.0801	102.3	8.556	2.037	28.82	5.315
—	—	—	.0788	99.10	8.830	2.0	27.93	5.480
*13			.0720	82.74	10.58	1.828	23.33	6.571
*14			.0641	65.63	13.33	1.628	18.50	8.285
*16			.0508	41.28	21.20	1.291	11.63	13.17
*17			.0453	32.74	26.74	1.150	9.23	16.61
*18			.0403	25.98	33.71	1.024	7.32	20.95
*19			.0359	20.58	42.51	.912	5.802	26.42
*22			.0253	10.27	85.24	.644	2.894	52.96
*24			.0201	6.46	135.5	.511	1.820	84.21
*26			.0159	4.06	215.5	.405	1.145	133.9
*27			.0142	3.22	271.7	.361	.908	168.9
*28			.0126	2.56	342.7	.321	.720	212.9

* When used in cable, weight and resistance of wire should be increased about 3% to allow for increase due to twist.

TABLE IX—STANDARD STRANDED COPPER CONDUCTORS
American wire gauge

circular mils	size AWG	number of wires	individual wire diam inches	cable diam inches	area square inches	weight lbs per 1000 ft	weight lbs per mile	maximum resistance ohms/1000 ft at 20° C
211,600	4/0	19	.1055	.528	0.1662	653.3	3,450	0.05093
167,800	3/0	19	.0940	.470	0.1318	518.1	2,736	0.06422
133,100	2/0	19	.0837	.419	0.1045	410.9	2,170	0.08097
105,500	1/0	19	.0745	.373	0.08286	325.7	1,720	0.1022
83,690	1	19	.0664	.332	0.06573	258.4	1,364	0.1288
66,370	2	7	.0974	.292	0.05213	204.9	1,082	0.1624
52,640	3	7	.0867	.260	0.04134	162.5	858.0	0.2048
41,740	4	7	.0772	.232	0.03278	128.9	680.5	0.2582
33,100	5	7	.0688	.206	0.02600	102.2	539.6	0.3256
26,250	6	7	.0612	.184	0.02062	81.05	427.9	0.4105
20,820	7	7	.0545	.164	0.01635	64.28	339.4	0.5176
16,510	8	7	.0486	.146	0.01297	50.98	269.1	0.6528
13,090	9	7	.0432	.130	0.01028	40.42	213.4	0.8233
10,380	10	7	.0385	.116	0.008152	32.05	169.2	1.038
6,530	12	7	.0305	.0915	0.005129	20.16	106.5	1.650
4,107	14	7	.0242	.0726	0.003226	12.68	66.95	2.624
2,583	16	7	.0192	.0576	0.002029	7.975	42.11	4.172
1,624	18	7	.0152	.0456	0.001275	5.014	26.47	6.636
1,022	20	7	.0121	.0363	0.008027	3.155	16.66	10.54

TABLE X—STANDARD COLOR CODE FOR RESISTORS AND CAPACITORS

Color	First Significant Figure (First Ring or Body Color)	Second Significant Figure (Second Ring or End Color)	Decimal Multiplier
Black	0	0	None
Brown	1	1	0
Red	2	2	00
Orange	3	3	,000
Yellow	4	4	0,000
Green	5	5	00,000
Blue	6	6	,000,000
Violet	7	7	0,000,000
Gray	8	8	00,000,000
White	9	9	000,000,000

Color-coded resistors and capacitors can be identified by means of this standard color code. Table XI, p. 174 shows the application of the code to various types of resistors, capacitors and codings used. An axial-lead composition resistor, for example, having yellow, violet, and yellow rings (in sequence from one end) has a resistance of 470,000 (470 K) ohms, since the first significant figure is 4 (first ring), the second significant figure is 7 (second ring) and the decimal multiplier (number of zeroes) is 4 (third ring). Similarly a molded mica capacitor, having orange, green and brown dots has a value of 350 $\mu\mu$f (micromicrofarads being understood). A fourth color band on resistors indicates the tolerance rating: gold = 5%, silver = 10%. The tolerance of resistors without a fourth band is 20%. The color code also determines the tolerance rating of capacitors. Thus, brown = 1%, red = 2%, green = 5%, etc. The voltage rating (working voltage) of capacitors is obtained by multiplying the numerical value of the color by 100. For example, yellow = 4 × 100 = 400 volts, green = 5 × 100 = 500 volts, blue = 6 × 100 = 600 volts, etc.

TABLE XI—COMPOSITION RESISTORS

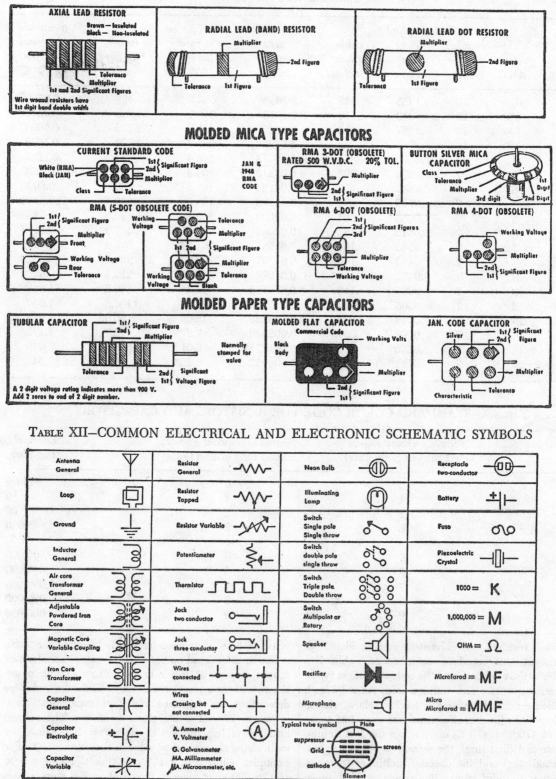

MOLDED MICA TYPE CAPACITORS

MOLDED PAPER TYPE CAPACITORS

TABLE XII—COMMON ELECTRICAL AND ELECTRONIC SCHEMATIC SYMBOLS

(Courtesy, The Institute of Radio Engineers)

APPENDIX II

MATHEMATICAL TABLES

TABLE XIII—COMMON LOGARITHMS OF NUMBERS *

N	0	1	2	3	4	5	6	7	8	9
10	0000	0043	0086	0128	0170	0212	0253	0294	0334	0374
11	0414	0453	0492	0531	0569	0607	0645	0682	0719	0755
12	0792	0828	0864	0899	0934	0969	1004	1038	1072	1106
13	1139	1173	1206	1239	1271	1303	1335	1367	1399	1430
14	1461	1492	1523	1553	1584	1614	1644	1673	1703	1732
15	1761	1790	1818	1847	1875	1903	1931	1959	1987	2014
16	2041	2068	2095	2122	2148	2175	2201	2227	2253	2279
17	2304	2330	2355	2380	2405	2430	2455	2480	2504	2529
18	2553	2577	2601	2625	2648	2672	2695	2718	2742	2765
19	2788	2810	2833	2856	2878	2900	2923	2945	2967	2989
20	3010	3032	3054	3075	3096	3118	3139	3160	3181	3201
21	3222	3243	3263	3284	3304	3324	3345	3365	3385	3404
22	3424	3444	3464	3483	3502	3522	3541	3560	3579	3598
23	3617	3636	3655	3674	3692	3711	3729	3747	3766	3784
24	3802	3820	3838	3856	3874	3892	3909	3927	3945	3962
25	3979	3997	4014	4031	4048	4065	4082	4099	4116	4133
26	4150	4166	4183	4200	4216	4232	4249	4265	4281	4298
27	4314	4330	4346	4362	4378	4393	4409	4425	4440	4456
28	4472	4487	4502	4518	4533	4548	4564	4579	4594	4609
29	4624	4639	4654	4669	4683	4698	4713	4728	4742	4757
30	4771	4786	4800	4814	4829	4843	4857	4871	4886	4900
31	4914	4928	4942	4955	4969	4983	4997	5011	5024	5038
32	5051	5065	5079	5092	5105	5119	5132	5145	5159	5172
33	5185	5198	5211	5224	5237	5250	5263	5276	5289	5302
34	5315	5328	5340	5353	5366	5378	5391	5403	5416	5428
35	5441	5453	5465	5478	5490	5502	5514	5527	5539	5551
36	5563	5575	5587	5599	5611	5623	5635	5647	5658	5670
37	5682	5694	5705	5717	5729	5740	5752	5763	5775	5786
38	5798	5809	5821	5832	5843	5855	5866	5877	5888	5899
39	5911	5922	5933	5944	5955	5966	5977	5988	5999	6010
40	6021	6031	6042	6053	6064	6075	6085	6096	6107	6117
41	6128	6138	6149	6160	6170	6180	6191	6201	6212	6222
42	6232	6243	6253	6263	6274	6284	6294	6304	6314	6325
43	6335	6345	6355	6365	6375	6385	6395	6405	6415	6425
44	6435	6444	6454	6464	6474	6484	6493	6503	6513	6522
45	6532	6542	6551	6561	6571	6580	6590	6599	6609	6618
46	6628	6637	6646	6656	6665	6675	6684	6693	6702	6712
47	6721	6730	6739	6749	6758	6767	6776	6785	6794	6803
48	6812	6821	6830	6839	6848	6857	6866	6875	6884	6893
49	6902	6911	6920	6928	6937	6946	6955	6964	6972	6981
50	6990	6998	7007	7016	7024	7033	7042	7050	7059	7067
51	7076	7084	7093	7101	7110	7118	7126	7135	7143	7152
52	7160	7168	7177	7185	7193	7202	7210	7218	7226	7235
53	7243	7251	7259	7267	7275	7284	7292	7300	7308	7316
54	7324	7332	7340	7348	7356	7364	7372	7380	7388	7396

* This table gives the mantissas of numbers, with the decimal point omitted. The characteristics are determined by inspection from the numbers. Antilogarithms may be found by looking up the numbers corresponding to the mantissa and then applying the characteristic to determine the decimal point.

COMMON LOGARITHMS OF NUMBERS—*Continued*

N	0	1	2	3	4	5	6	7	8	9
55	7404	7412	7419	7427	7435	7443	7451	7459	7466	7474
56	7482	7490	7497	7505	7513	7520	7528	7536	7543	7551
57	7559	7566	7574	7582	7589	7597	7604	7612	7619	7627
58	7634	7642	7649	7657	7664	7672	7679	7686	7694	7701
59	7709	7716	7723	7731	7738	7745	7752	7760	7767	7774
60	7782	7789	7796	7803	7810	7818	7825	7832	7839	7846
61	7853	7860	7868	7875	7882	7889	7896	7903	7910	7917
62	7924	7931	7938	7945	7952	7959	7966	7973	7980	7987
63	7993	8000	8007	8014	8021	8028	8035	8041	8048	8055
64	8062	8069	8075	8082	8089	8096	8102	8109	8116	8122
65	8129	8136	8142	8149	8156	8162	8169	8176	8182	8189
66	8195	8202	8209	8215	8222	8228	8235	8241	8248	8254
67	8261	8267	8274	8280	8287	8293	8299	8306	8312	8319
68	8325	8331	8338	8344	8351	8357	8363	8370	8376	8382
69	8388	8395	8401	8407	8414	8420	8426	8432	8439	8445
70	8451	8457	8463	8470	8476	8482	8488	8494	8500	8506
71	8513	8519	8525	8531	8537	8543	8549	8555	8561	8567
72	8573	8579	8585	8591	8597	8603	8609	8615	8621	8627
73	8633	8639	8645	8651	8657	8663	8669	8675	8681	8686
74	8692	8698	8704	8710	8716	8722	8727	8733	8739	8745
75	8751	8756	8762	8768	8774	8779	8785	8791	8797	8802
76	8808	8814	8820	8825	8831	8837	8842	8848	8854	8859
77	8865	8871	8876	8882	8887	8893	8899	8904	8910	8915
78	8921	8927	8932	8938	8943	8949	8954	8960	8965	8971
79	8976	8982	8987	8993	8998	9004	9009	9015	9020	9025
80	9031	9036	9042	9047	9053	9058	9063	9069	9074	9079
81	9085	9090	9096	9101	9106	9112	9117	9122	9128	9133
82	9138	9143	9149	9154	9159	9165	9170	9175	9180	9186
83	9191	9196	9201	9206	9212	9217	9222	9227	9232	9238
84	9243	9248	9253	9258	9263	9269	9274	9279	9284	9289
85	9294	9299	9304	9309	9315	9320	9325	9330	9335	9340
86	9345	9350	9355	9360	9365	9370	9375	9380	9385	9390
87	9395	9400	9405	9410	9415	9420	9425	9430	9435	9440
88	9445	9450	9455	9460	9465	9469	9474	9479	9484	9489
89	9494	9499	9504	9509	9513	9518	9523	9528	9533	9538
90	9542	9547	9552	9557	9562	9566	9571	9576	9581	9586
91	9590	9595	9600	9605	9609	9614	9619	9624	9628	9633
92	9638	9643	9647	9652	9657	9661	9666	9671	9675	9680
93	9685	9689	9694	9699	9703	9708	9713	9717	9722	9727
94	9731	9736	9741	9745	9750	9754	9759	9763	9768	9773
95	9777	9782	9786	9791	9795	9800	9805	9809	9814	9818
96	9823	9827	9832	9836	9841	9845	9850	9854	9859	9863
97	9868	9872	9877	9881	9886	9890	9898	9899	9903	9908
98	9912	9917	9921	9926	9930	9934	9939	9943	9948	9952
99	9956	9961	9965	9969	9974	9978	9983	9987	9991	9996

TABLE XIV
TABLE OF NATURAL TRIGONOMETRIC FUNCTIONS

Degrees	Sin	Cos	Tan	Cot	Sec	Csc	
0° 00′	.0000	1.0000	.0000	—	1.000	—	90° 00′
10	029	000	029	343.8	000	343.8	50
20	058	000	058	171.9	000	171.9	40
30	.0087	1.0000	.0087	114.6	1.000	114.6	30
40	116	9999	116	85.94	000	85.95	20
50	145	999	145	68.75	000	68.76	10
1° 00′	.0175	.9998	.0175	57.29	1.000	57.30	89° 00′
10	204	998	204	49.10	000	49.11	50
20	233	997	233	42.96	000	42.98	40
30	.0262	.9997	.0262	38.19	1.000	38.20	30
40	291	996	291	34.37	000	34.38	20
50	320	995	320	31.24	001	31.26	10
2° 00′	.0349	.9994	.0349	28.64	1.001	28.65	88° 00′
10	378	993	378	26.43	001	26.45	50
20	407	992	407	24.54	001	24.56	40
30	.0436	.9990	.0437	22.90	1.001	22.93	30
40	465	989	466	21.47	001	21.49	20
50	494	988	495	20.21	001	20.23	10
3° 00′	.0523	.9986	.0524	19.08	1.001	19.11	87° 00′
10	552	985	553	18.07	002	18.10	50
20	581	983	582	17.17	002	17.20	40
30	.0610	.9981	.0612	16.35	1.002	16.38	30
40	640	980	641	15.60	002	15.64	20
50	669	978	670	14.92	002	14.96	10
4° 00	.0698	.9976	.0699	14.30	1.002	14.34	86° 00′
10	727	974	729	13.73	003	13.76	50
20	756	971	758	13.20	003	13.23	40
30	.0785	.9969	.0787	12.71	1.003	12.75	30
40	814	967	816	12.25	003	12.29	20
50	843	964	846	11.83	004	11.87	10
5° 00	.0872	.9962	.0875	11.43	1.004	11.47	85° 00′
10	901	959	904	11.06	004	11.10	50
20	929	957	934	10.71	004	10.76	40
30	.0958	.9954	.0963	10.39	1.005	10.43	30
40	987	951	992	10.08	005	10.13	20
50	.1016	948	.1022	9.788	005	9.839	10
6° 00′	.1045	.9945	.1051	9.514	1.006	9.567	84° 00′
10	074	942	080	9.255	006	9.309	50
20	103	939	110	9.010	006	9.065	40
30	.1132	.9936	.1139	8.777	1.006	8.834	30
40	161	932	169	8.556	007	8.614	20
50	190	929	198	8.345	007	8.405	10
7° 00	.1219	.9925	.1228	8.144	1.008	8.206	83° 00′
10	248	922	257	7.953	008	8.016	50
20	276	918	287	7.770	008	7.834	40
30	.1305	.9914	.1317	7.596	1.009	7.661	30
40	334	911	346	7.429	009	7.496	20
50	363	907	376	7.269	009	7.337	10
8° 00′	.1392	.9903	.1405	7.115	1.010	7.185	82° 00′
10	421	899	435	6.968	010	7.040	50
20	449	894	465	6.827	011	6.900	40
30	.1478	.9890	.1495	6.691	1.011	6.765	30
40	507	886	524	6.561	012	6.636	20
50	536	881	554	6.435	012	6.512	10
9° 00′	.1564	.9877	.1584	6.314	1.012	6.392	81° 00
	Cos	Sin	Cot	Tan	Csc	Sec	Degrees

TABLE OF NATURAL TRIGONOMETRIC FUNCTIONS—*Continued*

Degrees	Sin	Cos	Tan	Cot	Sec	Csc	
9° 00′	.1564	.9877	.1584	6.314	1.012	6.392	81° 00′
10	593	872	614	197	013	277	50
20	622	868	644	084	013	166	40
30	.1650	.9863	.1673	5.976	1.014	6.059	30
40	679	858	703	871	014	5.955	20
50	708	853	733	769	015	855	10
10° 00′	.1736	.9848	.1763	5.671	1.015	5.759	80° 00
10	765	843	793	576	016	665	50
20	794	838	823	485	016	575	40
30	.1822	.9833	.1853	5.396	1.017	5.487	30
40	851	827	883	309	018	403	20
50	880	822	914	226	018	320	10
11° 00	.1908	9816	.1944	5.145	1.019	5.241	79° 00′
10	937	811	974	066	019	164	50
20	965	805	.2004	4.989	020	089	40
30	.1994	.9799	.2035	4.915	1.020	5.016	30
40	.2022	793	065	843	021	4.945	20
50	051	787	095	773	022	876	10
12° 00′	.2079	.9781	.2126	4.705	1.022	4.810	78° 00′
10	108	775	156	638	023	745	50
20	136	769	186	574	024	682	40
30	.2164	.9763	.2217	4.511	1.024	4.620	30
40	193	757	247	449	025	560	20
50	221	750	278	390	026	502	10
13° 00′	.2250	.9744	.2309	4.331	1.026	4.445	77° 00′
10	278	737	339	275	027	390	50
20	306	730	370	219	028	336	40
30	.2334	.9724	.2401	4.165	1.028	4.284	30
40	363	717	432	113	029	232	20
50	391	710	462	061	030	182	10
14° 00′	.2419	.9703	.2493	4.011	1.031	4.134	76° 00′
10	447	696	524	3.962	031	086	50
20	476	689	555	914	032	039	40
30	.2504	.9681	.2586	3.867	1.033	3.994	30
40	532	674	617	821	034	950	20
50	560	667	648	776	034	906	10
15° 00′	.2588	.9659	.2679	3.732	1.035	3.864	75° 00′
10	616	652	711	689	036	822	50
20	644	644	742	647	037	782	40
30	.2672	.9636	.2773	3.606	1.038	3.742	30
40	700	628	805	566	039	703	20
50	728	621	836	526	039	665	10
16° 00′	.2756	.9613	.2867	3.487	1.040	3.628	74° 00′
10	784	605	899	450	041	592	50
20	812	596	931	412	042	556	40
30	.2840	.9588	.2962	3.376	1.043	3.521	30
40	868	580	994	340	044	487	20
50	896	572	.3026	305	045	453	10
17° 00′	.2924	.9563	.3057	3.271	1.046	3.420	73° 00′
10	952	555	089	237	047	388	50
20	979	546	121	204	048	356	40
30	.3007	.9537	.3153	3.172	1.049	3.326	30
40	035	528	185	140	049	295	20
50	062	520	217	108	050	265	10
18° 00′	.3090	.9511	.3249	3.078	1.051	3.236	72° 00′
	Cos	Sin	Cot	Tan	Csc	Sec	Degrees

TABLE OF NATURAL TRIGONOMETRIC FUNCTIONS—*Continued*

Degrees	Sin	Cos	Tan	Cot	Sec	Csc	
18° 00′	.3090	.9511	.3249	3.078	1.051	3.236	72° 00′
10	118	502	281	047	052	207	50
20	145	492	314	018	053	179	40
30	.3173	.9483	.3346	2.989	1.054	3.152	30
40	201	474	378	960	056	124	20
50	228	465	411	932	057	098	10
19° 00′	.3256	.9455	.3443	2.904	1.058	3.072	71° 00′
10	283	446	476	877	059	046	50
20	311	436	508	850	060	021	40
30	.3338	.9426	.3541	2.824	1.061	2.996	30
40	365	417	574	798	062	971	20
50	393	407	607	773	063	947	10
20° 00′	.3420	.9397	.3640	2.747	1.064	2.924	70° 00′
10	448	387	673	723	065	901	50
20	475	377	706	699	066	878	40
30	.3502	.9367	.3739	2.675	1.068	2.855	30
40	529	356	772	651	069	833	20
50	557	346	805	628	070	812	10
21° 00′	.3584	.9336	.3839	2.605	1.071	2.790	69° 00′
10	611	325	872	583	072	769	50
20	638	315	906	560	074	749	40
30	.3665	.9304	.3939	2.539	1.075	2.729	30
40	692	293	973	517	076	709	20
50	719	283	.4006	496	077	689	10
22° 00′	.3746	.9272	.4040	2.475	1.079	2.669	68° 00′
10	773	261	074	455	080	650	50
20	800	250	108	434	081	632	40
30	.3827	.9239	.4142	2.414	1.082	2.613	30
40	854	228	176	394	084	595	20
50	881	216	210	375	085	577	10
23° 00′	.3907	.9205	.4245	2.356	1.086	2.559	67° 00′
10	934	194	279	337	088	542	50
20	961	182	314	318	089	525	40
30	.3987	.9171	.4348	2.300	1.090	2.508	30
40	.4014	159	383	282	092	491	20
50	041	147	417	264	093	475	10
24° 00′	.4067	.9135	.4452	2.246	1.095	2.459	66° 00′
10	094	124	487	229	096	443	50
20	120	112	522	211	097	427	40
30	.4147	.9100	.4557	2.194	1.099	2.411	30
40	173	088	592	177	100	396	20
50	200	075	628	161	102	381	10
25° 00′	.4226	.9063	.4663	2.145	1.103	2.366	65° 00′
10	253	051	699	128	105	352	50
20	279	038	734	112	106	337	40
30	.4305	.9026	.4770	2.097	1.108	2.323	30
40	331	013	806	081	109	309	20
50	358	001	841	066	111	295	10
26° 00′	.4384	.8988	.4877	2.050	1.113	2.281	64° 00′
10	410	975	913	035	114	268	50
20	436	962	950	020	116	254	40
30	.4462	.8949	.4986	2.006	1.117	2.241	30
40	488	936	.5022	1.991	119	228	20
50	514	923	059	977	121	215	10
27° 00′	.4540	.8910	.5095	1.963	1.122	2.203	63° 00′
	Cos	Sin	Cot	Tan	Csc	Sec	Degrees

TABLE OF NATURAL TRIGONOMETRIC FUNCTIONS—*Continued*

Degrees	Sin	Cos	Tan	Cot	Sec	Csc	
27° 00′	.4540	.8910	.5095	1.963	1.122	2.203	63° 00′
10	566	897	132	949	124	190	50
20	592	884	169	935	126	178	40
30	.4617	.8870	.5206	1.921	1.127	2.166	30
40	643	857	243	907	129	154	20
50	669	843	280	894	131	142	10
28° 00′	.4695	.8829	.5317	1.881	1.133	2.130	62° 00′
10	720	816	354	868	134	118	50
20	746	802	392	855	136	107	40
30	.4772	.8788	.5430	1.842	1.138	2.096	30
40	797	774	467	829	140	085	20
50	823	760	505	816	142	074	10
29° 00′	.4848	8746	.5543	1.804	1.143	2.063	61° 00′
10	874	732	581	792	145	052	50
20	899	718	619	780	147	041	40
30	.4924	.8704	.5658	1.767	1.149	2.031	30
40	950	689	696	756	151	020	20
50	975	675	735	744	153	010	10
30° 00′	.5000	.8660	.5774	1.732	1.155	2.000	60° 00′
10	025	646	812	720	157	1.990	50
20	050	631	851	709	159	980	40
30	.5075	.8616	.5890	1.698	1.161	1.970	30
40	100	601	930	686	163	961	20
50	125	587	969	675	165	951	10
31° 00′	.5150	.8572	.6009	1.664	1.167	1.942	59° 00′
10	175	557	048	653	169	932	50
20	200	542	088	643	171	923	40
30	.5225	.8526	.6128	1.632	1.173	1.914	30
40	250	511	168	621	175	905	20
50	275	496	208	611	177	896	10
32° 00′	.5299	.8480	.6249	1.600	1.179	1.887	58° 00′
10	324	465	289	590	181	878	50
20	348	450	330	580	184	870	40
30	.5373	.8434	.6371	1.570	1.186	1.861	30
40	398	418	412	560	188	853	20
50	422	403	453	550	190	844	10
33° 00′	.5446	.8387	.6494	1.540	1.192	1.836	57° 00′
10	471	371	536	530	195	828	50
20	495	355	577	520	197	820	40
30	.5519	.8339	.6619	1.511	1.199	1.812	30
40	544	323	661	501	202	804	20
50	568	307	703	1.492	204	796	10
34° 00′	.5592	.8290	.6745	1.483	1.206	1.788	56° 00′
10	616	274	787	473	209	781	50
20	640	258	830	464	211	773	40
30	.5664	.8241	.6873	1.455	1.213	1.766	30
40	688	225	916	446	216	758	20
50	712	208	959	437	218	751	10
35° 00′	.5736	.8192	.7002	1.428	1.221	1.743	55° 00′
10	760	175	046	419	223	736	50
20	783	158	089	411	226	729	40
30	.5807	.8141	.7133	1.402	1.228	1.722	30
40	831	124	177	393	231	715	20
50	854	107	221	385	233	708	10
36° 00′	.5878	.8090	.7265	1.376	1.236	1.701	54° 00′
	Cos	Sin	Cot	Tan	Csc	Sec	Degrees

TABLE OF NATURAL TRIGONOMETRIC FUNCTIONS—*Continued*

Degrees	Sin	Cos	Tan	Cot	Sec	Csc	
36° 00′	.5878	.8090	.7265	1.376	1.236	1.701	54° 00′
10	901	073	310	368	239	695	50
20	925	056	355	360	241	688	40
30	.5948	.8039	7400	1.351	1.244	1.681	30
40	972	021	445	343	247	675	20
50	995	004	490	335	249	668	10
37° 00′	.6018	.7986	.7536	1.327	1.252	1.662	53° 00′
10	041	969	581	319	255	655	50
20	065	951	627	311	258	649	40
30	.6088	.7934	.7673	1.303	1.260	1.643	30
40	111	916	720	295	263	636	20
50	134	898	766	288	266	630	10
38° 00′	.6157	.7880	.7813	1.280	1.269	1.624	52° 00′
10	180	862	860	272	272	618	50
20	202	844	907	265	275	612	40
30	.6225	.7826	.7954	1.257	1.278	1.606	30
40	248	808	.8002	250	281	601	20
50	271	790	050	242	284	595	10
39° 00′	.6293	.7771	8098	1.235	1.287	1.589	51° 00′
10	316	753	146	228	290	583	50
20	338	735	195	220	293	578	40
30	.6361	.7716	.8243	1.213	1.296	1.572	30
40	383	698	292	206	299	567	20
50	406	679	342	199	302	561	10
40° 00′	.6428	.7660	.8391	1.192	1.305	1.556	50° 00′
10	450	642	441	185	309	550	50
20	472	623	491	178	312	545	40
30	.6494	.7604	.8541	1.171	1.315	1.540	30
40	517	585	591	164	318	535	20
50	539	566	642	157	322	529	10
41° 00′	.6561	.7547	.8693	1.150	1.325	1.524	49° 00′
10	583	528	744	144	328	519	50
20	604	509	796	137	332	514	40
30	.6626	.7490	.8847	1.130	1.335	1.509	30
40	648	470	899	124	339	504	20
50	670	451	952	117	342	499	10
42° 00′	.6691	.7431	.9004	1.111	1.346	1.494	48° 00′
10	713	412	057	104	349	490	50
20	734	392	110	098	353	485	40
30	.6756	.7373	.9163	1.091	1.356	1.480	30
40	777	353	217	085	360	476	20
50	799	333	271	079	364	471	10
43° 00′	.6820	.7314	.9325	1.072	1.367	1.466	47° 00′
10	841	294	380	066	371	462	50
20	862	274	435	060	375	457	40
30	.6884	.7254	.9490	1.054	1.379	1.453	30
40	905	234	545	048	382	448	20
50	926	214	601	042	386	444	10
44° 00′	.6947	.7193	9657	1.036	1.390	1.440	46° 00′
10	967	173	713	030	394	435	50
20	988	153	770	024	398	431	40
30	.7009	.7133	.9827	1.018	1.402	1.427	30
40	030	112	884	012	406	423	20
50	050	092	942	006	410	418	10
45° 00′	.7071	.7071	1.0000	1.000	1.414	1.414	45° 00′
	Cos	Sin	Cot	Tan	Csc	Sec	Degrees

ANSWERS

Exercise No. 1

1. No Answer.
2. 92 in nature; 6 to 8 artificially produced.
3. No Answer.
4. No Answer.
5. No Answer.
6. 10 neutrons. Element is inert, because outer shell is filled.
7. When two electrons are knocked out from the atom, *double ionization* is said to occur.
8. Free electrons are *not bound* to the atom.
9. Conductors—many free electrons; semiconductors—relatively few free electrons; insulators—practically no free electrons.
10. A flow of free electrons towards positive terminal of supply.
11. Just as many electrons leave the wire at one end than enter it at the other end; hence, wire remains electrically neutral.
12. No Answer.
13. (a) mechanical; (b) chemical; (c) photoelectric; (d) thermoelectric; (e) piezoelectric. (See Summary of chapter 1.)

Exercise No. 2

1. No Answer.
2. First a charge of opposite polarity is induced, converging the leaves; then a charge of the same polarity is transferred by contact, which diverges the leaves.
3. First opposite charges are induced, attracting the balls. After contact, like charges repel the balls.
4. (a) contact the sphere with a positive charge till it is charged.
   (b) Bring a negative charge near the sphere; then momentarily ground the sphere and remove the external negative charge, thus freeing the bound positive charge.
5. No Answer.
6. 30 dynes/esu.
7. 50 dynes *attraction*.
8. (a) 1200 esu; (b) 1.33 dynes/esu.
9. (a) 8 esu; (b) 16 esu.
10. 6000 joules.
11. 4 V/in., 48 V/ft.
12. (a) 50 μf; (b) 500 in. $\times$ 500 in.

Exercise No. 3

1. (a) Place it near the north pole of the marked magnet; the pole that is repelled is the north pole; (b) suspend the magnet and identify the north-seeking pole.
2. Any ferromagnetic material will attract the poles of a magnet, but only the poles of another magnet will repel like poles.
3. The right end is the north pole.
4. The bar has been stroked with the south pole of a permanent magnet from the center outward, first in one direction, then in the other. This produces *consecutive* north poles at the ends.
5. 48 e.m.u.
6. 12,280 dynes repulsion.
7. (a) 4 oersteds; (b) 15 e.m.u.
8. (a) 500 gauss; (b) 72,000 maxwells.

Exercise No. 4

1. No Answer.
2. The thermal emf is *directly* affected by the temperature; the thermoelectric current is also affected by the *resistance* of the wire, which—in turn—depends on the temperature.
3. Increasing the temperature difference increases the emf; increasing the area of the metals in contact (or decreasing their resistance) will increase the current capacity.
4. No Answer.
5. No Answer.
6. The photoemissive cell requires amplification; the others do not.
7. No Answer.
8. The deformation of a piezoelectric crystal when an emf is connected to its faces.
9. Time the echo returning from the opposite side of the metal or from the ocean floor. Knowing the speed of sound in either medium, the distance traveled (thickness) can be computed.

Exercise No. 5

1. The penny and the dime are made of different metals; when in contact with the moist tongue, an emf is generated by chemical action. Ion flow in the electrolyte (tongue) creates the sour taste.

2. *Open circuit:* Zinc ionizes, $Z_n \longrightarrow Zn^{++} + 2e^-$
Copper ionizes slightly, $Cu \longrightarrow Cu^{++} + 2e^-$
Sulfuric acid dissociates,
$$H_2SO_4 \longrightarrow 2H^+ + SO_4^{--}$$
Closed circuit: Equilibrium is upset and the surplus of electrons on the zinc electrode flows through the wire to the copper electrode. More zinc dissolves and the additional Zn^{++} ions displace H^+ ions and drive them over to the copper electrode. There neutral hydrogen gas is evolved by the reaction: $H^+ + e^- \longrightarrow H°$

3. (a) 2.52 volts; (b) 2.01 volts.

4. No Answer.

5. Displaced hydrogen ions combine with electrons arriving at the positive electrode (see 11 above), thus evolving hydrogen gas, which coats the electrode. This polarization reduces the emf of the cell, unless a *depolarizer* removes the hydrogen from the electrode.

6. It decreases it, thus increasing the emf.

7. No Answer.

8. Connect 10 cells in series.

9. Connect them in parallel.

10. Salts, bases, and acids are electrolytes.

11. *Cathode:* $2H^+ + 2e^- \longrightarrow H_2 \uparrow$ (hydrogen gas)
Anode: $2SO_4^{--} +$
$$2H_2O \longrightarrow \underbrace{4H^+ + 2SO_4^{--}}_{2H_2SO_4} + 4e^- + O_2 \uparrow$$

12. Make either object the cathode in a salt solution of the covering metal (silver or gold); make the anode pure silver or pure gold as required. Cannot nickel-plate a silver spoon, since silver is *less active* than nickel.

13. No Answer.

14. $\dfrac{15 \times 6}{5 \times 4} \times 15 = 67.5$ grams.

15. (a) 0.0003 gm/coulomb; (b) 15 grams.

16. No Answer.

17. $PbO_2 + Pb + 2H_2SO_4 \xrightarrow[\xleftarrow[\text{charge}]{}]{\text{discharge}} 2PbSO_4 + H_2O$
Pos. Plate Neg. Plate

Exercise No. 6

1. 1,800 coulombs.

2. 0.000005 volt; 0.015 amp; 2,500,000 ohms.

3. 1 absolute ampere = 1.000165 international ampere.

4. No Answer.

5. 1 international ohm = 1.000495 absolute ohms.

6. Remains the same.

7. 0.204 inch.

8. 0.462 ohm.

9. 0.299 ohm.

10. 0.397 ohm.

11. 2720°C.

12. 271 ohms.

13. 0.3 ampere.

14. 220 volts.

15. No.

16. (a) 5.9 volts; (b) 2.95 ohms.

17. 0.0735 ohm.

Exercise No. 7

1. In a *series circuit* the current is everywhere the same and the sum of the voltage drops must equal the emf of the source. A series circuit either operates throughout all parts, or not at all.

2. 132 volts emf, 24 V, 48 V, 60 V.

3. 0.75 amp line current, 112.5 volts across lamp, 7.5 volts drop in the lines.

4. 0.5 amp through both.

5. 2 ohms.

6. In a *parallel circuit* the total current divides into a number of separate branches; the total current flowing in and out of the junction points must equal the sum of the branch currents. The voltage drop across each branch is the same and equals the emf of the source. Failure in a parallel branch does *not* disable the remainder of the circuit.

7. $E = I_1 R_1 = I_2 R_2$.

8. 4.5 amps through 20-ohm branch; 2.25 amps through 40-ohm branch; 90 volys emf, 9.24 ohms total resistance; 9.75 amps total current.

9. 48 ohms.

10. 10 amps, 5 amps, 15 amps total.

11. 0.14 amp.

12. 94.3 V and 125.7 V.

13. 5 ohms.

14. No Answer.

15. $R_x = \dfrac{R_a}{R_b} \times R_s$, as before.

Exercise No. 8

1. Iron for *series* connection; copper for *parallel* connection.

2. 1,720,000 calories.

3. 722,000 cals, 4.2¢/hour.

4. $2.27.
5. 876 watts.
6. $1,640.00.
7. 5 ohms; 1125 watts.

Exercise No. 9

1. Right to left.
2. The sum, or 0.63 oersted.
3. Same direction: midway between wires; opposite direction: a point equal to half the distance between the wires, beyond either wire.
4. 79.5 amps.
5. One ampere is that current, which in a circular loop of 1 cm. radius will produce a field intensity of $2\pi/10$ or 0.62831 oersteds at the center.
6. 10.5 oersteds.
7. 943 oersteds.
8. No Answer.
9. 6.28 oersteds.
10. No Answer.
11. No Answer.
12. The permeability of its core and the magnetizing force (ampere-turns). Ampere-turns are the magnetizing force expressed by the product of the number of turns in the winding and the strength of the current (amperes) flowing through it.
13. The permeability is decreasing because of hysteresis.
14. Soft iron is suitable for electromagnets because of its low hysteresis losses (small area of loop); hard iron is suitable for permanent magnets only.
15. Magnetic Flux $= \dfrac{\text{Magnetomotive Force}}{\text{Reluctance}} (\phi = \dfrac{F}{R})$
16. Reluctance $= \dfrac{\text{Length of Path}}{\text{Permeability} \times \text{Area}}$;

$$R_{Par} = \dfrac{1}{\dfrac{1}{R_1} + \dfrac{1}{R_2} + \dfrac{1}{R_3} + \cdots}$$

17. Divide by 1.2566.
18. It quadruples.
19. 5 dynes.
20. No Answer.

Exercise No. 10

1. Speed of motion, strength of magnet, number of turns on coil.
2. Motion toward or away from coil.

3. Yes.
4. 0.196 volt directed clockwise, looking down.
5. 2500 gauss.
6. 3 volts.
7. Conservation of Energy.
8. Yes.
9. A generator converts mechanical work into electric energy; a motor converts electric energy into into mechanical work. An a-c generator has slip rings; a d-c generator has a commutator.
10. The sine-wave output is converted into a series of positive, unidirectional alternations (pulsating d.c.).
11. No Answer.
12. The inductance of the coil opposes the interruption of the current through the energy stored in its magnetic field. This results in the spark.
13. The inductance of the coil opposes the buildup or change in the current, causing it to lag behind the voltage.
14. 6 henrys.
15. 0.667
16. The transformer is more efficient.
17. $\dfrac{Ep}{Es} = \dfrac{Is}{Ip} = \dfrac{Np}{Ns}$ (E = voltage; I = current; p = primary; s = secondary.)
18. The thin wires.
19. 20:1; 50 turns on the secondary.
20. Losses due to the resistance of the windings of the transformer (copper loss), hysteresis and eddy currents (iron losses).
21. Eddy currents are circulating currents induced in the iron masses of the transformer core; they can be reduced by laminating the core into thin strips or segments.
22. % Efficiency $= \dfrac{\text{Power Output}}{\text{Power Input}} \times 100$
23. 1 ampere.

Exercise No. 11

1. No Answer
2. (a) 2π; (b) ½, ⅛, 1 cycles. ¹⁄₁₂, ⅙, 2 cycles. ⅓, ¾, ¾-cycle.
3. f = 50 cps; e = 1115 sin 314 t.
4. (a) 21.2 amps; (b) 9.27 amps.
5. $I_{rms} = 0.707\ I_m$; $I_{av} = 0.636\ I_m$.
6. (a) 31.1 amps, 545 volts; (b) 19.8 amps, 346 volts.
7. Sine wave 2 leads wave 3 by 170 degrees.
8. No.

9. 390 mph in a north-westerly direction.
10. (a) 250 units, 53.2°; (b) Yes.
11. Reactance = 250 ohms; resistance = 433 ohms.

Exercise No. 12

1. They are in phase.
2. 8 amps, 0 angle, 1600 watts.
3. Diameter of wire, square root of frequency.
4. 1 henry, 107 millihenrys.
5. (a) 40 henrys; (b) 20 henrys.
6. Applied and cemf, 180° out of phase; current *lags* applied voltage by 90° and *leads* counter emf by 90°.
7. Proportional to f and L; $X_L = 2\pi f L$.
8. 628 ohms; .0159 ampere.
9. Generator *A* is a.c., *B* is d.c.; impedance = 50 ohms; d-c resistance = 2 ohms; inductive reactance = 49.9 ohms; power from *A* = 8 watts; power from *B* = 5 kilowatts.
10. Because it is continually charged and discharged, which results in a current oscillating to and fro in the connecting wires.
11. $X_c = \dfrac{1}{2\pi fC}$; capacitive reactance decreases with frequency; inductive reactance increases with frequency.
12. series: 2.4 μf; parallel: 10 μf.
13. $X_C = 318$ ohms; I = 63 milliamps.
14. Current leads voltage by 90°; no power consumed.
15. $Z = \sqrt{R^2 + (X_L - X_C)^2}$.
16. Net reactance = 120 ohms; impedance = 130 ohms; phase angle = 67.4°; the circuit is inductive.

Exercise No. 13

1. P.F. = $\dfrac{\text{Real Power}}{\text{Appar. Power}}$; 0-1; high P.F. is desirable.
2. Apparent power = EI; real power = EI cos θ; reactive power = E I sin θ.
3. App. power = 6000 volt-amps; P.F. = 0.667; true power = 4000 watts; reactive power = 4,450 vars.

4. (a) 37.7 ohms; (b) 41.7 ohms; (c) 5.28 amps; (d) 64.5°.
5. Impedance = 3770 ohms, eq. inductance = 10 henrys with 25 ohms winding resistance.
6. 1.06 amps, 58°, 33 μf, 56.2 watts, 90 vars.
7. $X_L = 125.6$ ohms; $X_C = 442.3$ ohms; Z = 374.6 ohms; I = 53.4 ma; I leads E by a phase angle of 57.7°; if capacitor is removed, I = 84.6 ma; if coil is removed, I = 41.2 ma.
8. $E_R = 146$ volts; $E_L = 150$ volts; $E_C = 219$ volts; Z = 151 ohms; $I_t = 1.46$ amps; $\theta = 22°$ (*I lags E*); app. Power = 322 volt-amps; true power = 298 watts.
9. $I_R = 0.12$ amp; $I_L = 0.117$ amp; $I_C = 0.08$ amp; $I_t = 0.168$ amp; Z = 71.5 ohms; power = 1.99 watts.

Exercise No. 14

1. No Answer
2. No Answer
3. (a) inertia of mass; (b) capacitance; (c) charge; (d) potential energy; (e) energy of magnetic field.
4. No Answer
5. When the impressed frequency equals the natural frequency of oscillations; i.e. at the resonant frequency.
6. $X_L = X_C$; $f_r = \dfrac{1}{2\pi\sqrt{LC}}$
7. No Answer.
8. 0.352 henry.
9. (a) $f_r = 796$ kc; (b) I = 25 ma; (c) and (d) $E_L = E_C = 20$ volts.
10. 40
11. 1. unity power factor; 2. $X_L = X_C$; 3. impedance a maximum, or line current a minimum.
$$f_r = \dfrac{1}{2\pi\sqrt{L\,C}}.$$
12. No Answer
13. (a) $Z = Q\,X_L$; (b) $Z = \dfrac{L}{CR}$; pure resistance, $\theta = 0$.
14. $f_r = 60$ cps; Z = R = 143 ohms.
15. (a) 417 $\mu\mu f$; (b) 47.2; (c) 40,000 ohms; (d) 0.25 ma; (e) 11.8 ma.